Multisim® Experiments

for

DC/AC, Digital, and Devices Courses

Gary D. Snyder and David M. Buchla

Prentice Hall

Boston   Columbus   Indianapolis   New York   San Francisco   Upper Saddle River

Amsterdam   Cape Town   Dubai   London   Madrid   Milan   Munich   Paris   Montreal   Toronto

Delhi   Mexico City   Sao Paulo   Sydney   Hong Kong   Seoul   Singapore   Taipei   Tokyo

Editorial Director: Vern Anthony
Acquisitions Editor: Wyatt Morris
Editorial Assistant: Tanika Henderson
Director of Marketing: David Gesell
Marketing Manager: Kara Clark
Senior Marketing Manager: Alicia Wozniak
Marketing Assistant: Les Roberts
Senior Managing Editor: JoEllen Gohr

Project Manager: Rex Davidson
Senior Operations Supervisor: Pat Tonneman
Operations Specialist: Laura Weaver
Art Director: Jayne Conte
Cover Image: Fotolia
Printer/Binder: Bind-Rite Graphics
Cover Printer: Lehigh-Phoenix

10 9 8 7 6 5 4 3 2 1

Prentice Hall
is an imprint of

www.pearsonhighered.com

ISBN-13: 978-0-13-211388-5
ISBN-10:　 0-13-211388-0

Preface

About This Workbook

As the field of electronics has changed, the processes and tools to design and evaluate electronic circuits have also changed. In the past, manufacturers would draft schematics by hand, build prototype circuits, and test the prototypes to evaluate how well the circuit design met the design requirements. In order to develop new products as quickly and inexpensively as possible, manufacturers now use design and simulation tools that allow them to design and evaluate circuits before going through the costly process of putting the design in hardware, and to easily incorporate and evaluate changes to the circuit.

The National Instruments Multisim® software is a versatile design and simulation program. The intent of this workbook is to simulate a laboratory experience in electronics and help you develop a working knowledge of the Multisim software to enter and analyze circuit designs. Features and advantages of this manual and the Multisim software are:

- The manual provides a good overview of dc/ac, digital, and devices topics.
- Pre-constructed circuit files minimize set-up time for experiments.
- Multisim allows students to work safely with components, circuits, and equipment that would be impractical or unavailable in most school labs.
- The virtual nature of experiments extends study time by allowing students to work at home as well as at school.
- Experiments and review questions permit and encourage independent study of electronic topics with Multisim.
- Instructors can use the manual experiments to reinforce and supplement practical labs.
- The manual organization and content are an excellent complement to Course Connect on-line dc/ac, digital, and devices courses.

The circuits in this manual illustrate fundamental concepts in dc/ac, digital, and device electronics. Each section will contain some background theory for the circuits that you will investigate, but only to help provide context for the specific topics that the section will cover. For best results, you should use this workbook to supplement, rather than replace, a textbook that discusses the subject material in depth. This manual provides suggested reading for each experiment. The content of this manual is designed to track the Pearson on-line Course Connect electronics lessons (DC/AC, Digital, and Devices), but it can also supplement any introductory electronics course.

Workbook Organization

This workbook consists of three sections, each of which contains a separate table of contents and fifteen experiments. The first section covers basic dc and ac electronics, the second section covers digital electronics, and the third section covers electronic devices. Each experiment consists of two or more parts and typically will consist of the following components.

Introduction

The Introduction section provides background material for the experiment and provides an overview of the experiment.

Reading

The Reading section provides additional references so that students may prepare for the experiment.

Key Objectives

The Key Objectives section summarizes the intent of each part of the experiment.

Multisim Files

The Multisim Files section specifies prepared Multisim circuit files that are associated with the experiment. These files are contained on the CD that is available with this manual.

Experiment Part Sections

These sections detail the tasks that students are to complete in performing the experiment. Each experiment typically consists of at least two parts.

Conclusions

The Conclusions section provides space for students to summarize their experimental results in their own words.

Questions

The Questions section provides an opportunity for students to review their experiments and encourages students to pursue further independent investigation with Multisim.

Using This Workbook

Using Windows

This workbook assumes that you are familiar with using the Microsoft® Windows® operating system (starting applications, opening files, etc.) and with standard Windows operations and terminology (double-clicking icons, minimizing windows, etc.). For simplicity, "click" is the same as "left-click" and means to click the left mouse button, and "double-click" means to click the left mouse button twice. If you have any questions about Windows, refer to the on-line help or standard Windows documentation.

Multisim Software

This workbook also requires that you have a copy of the Multisim software. The references to sample circuits are by filename only, but each file has a Multisim (.MSx) extension.

Workbook Conventions for Components

Italic font identifies quantities, components, and component values such as resistor R_1 or dc voltage supply V_S, or "$R_1 = 100$ ohms" or "$V_S = 12$ V".

Non-italic subscripts identify dc quantities, such as "$V_S = 10$ V$_{DC}$". Italic subscripts identify ac quantities, such as "$V_s = 10$ V$_{rms}$".

Although Multisim fonts do not support subscripts, standard convention is to subscript the numbers for reference designators (e.g., "R_1" rather than "$R1$"). For consistency, all component reference designators in this manual use non-italic subscripted numbers.

Circuit Restrictions

Some of the Multisim circuits have circuit restrictions to prevent students from seeing hidden component values or component faults. These restrictions are password protected.

Protoboarding Circuits

 WARNING

All Multisim circuits and experimental procedures
associated with this manual are for educational purposes
only. Several circuits show 120 Vac sources and could
develop potentially dangerous voltages in actual circuits.
Students should not attempt to construct, energize, or
operate these circuits with actual components without
observing all laboratory safety procedures and the
supervision of a qualified electronics instructor.

One of the most popular ways of constructing and testing electronics circuits is with a solderless
breadboard, also known as a protoboard. Protoboards provide a simple way to quickly connect leaded
electronic components without soldering. Because connections do not require soldering, protoboards allow
technicians and engineers to easily build, modify, and test prototype circuits to evaluate and verify the
operation of circuit designs. Several experiments in this manual show how circuits could be built using a
protoboard. The figure below shows an Experimentor 300 protoboard (Radio Shack P/N 276-174), which is
one popular protoboard model. Protoboards are available in various models from various manufacturers
and differ somewhat in specific features, but the basic construction and use are the same.

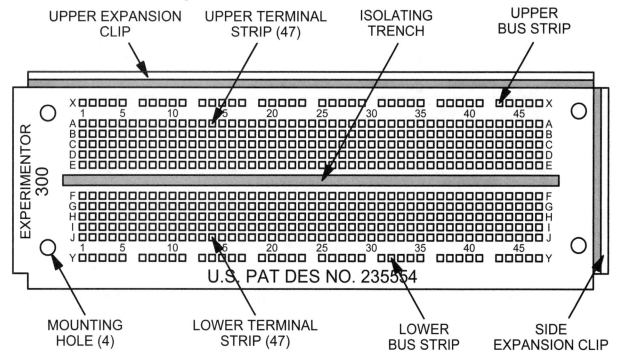

The Experimentor 300 Protoboard

As the figure above shows, the protoboard consists of a number of holes into which the leads of
components can be inserted. Conductive spring clips provide electrical connections between specific
groups of holes, called strips. There are two types of strips.

- *Bus strips* are horizontal connections that allow the protoboard to distribute supply voltages and
  ground references to all parts of the circuit. The Experimentor 300 has two bus strips (the X strip
  on the upper half and the Y strip on the lower half), but other protoboards can have several. Some

protoboards electrically isolate the right and left halves of the bus strips, but Experimentor 300 bus strips run the entire width of the board.

- *Terminal strips* are vertical connections that allow the protoboard to selectively connect components together. The number of terminal strips varies with the protoboard manufacturer and model. The Experimentor 300 has 47 terminal strips with five connections each (labeled A through E) on the upper half of the board and 47 terminal strips with five connections each (labeled F through J) on the lower half of the board. The upper and lower terminal strips are electrically isolated by an isolating trench in the center of the board.

In addition to the the bus and terminal strips, the Experimentor 300 has four mounting holes for attaching the protoboard to a board or other backing for greater robustness. The protoboard also has expansion clips so that users can mechanically connect two or more protoboards together vertically, horizontally, or both.

The figure below shows a circuit schematic and one way to connect the circuit on a protoboard. Note that this is not the only (or most efficient way) to connect the circuit on the protoboard and is intended to provide some practice in recognizing electrical connections on a protoboard.

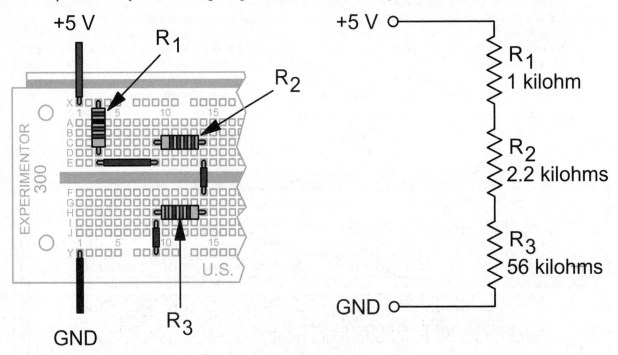

Sample Protoboard Circuit

As the figure above shows, the upper bus strip X electrically connects the wire from the +5 V supply to the top of resistor R_1. Upper terminal strip 3 electrically connects the bottom of R_1 to the left side of the first jumper wire. Upper terminal strip 9 electrically connects the right side of the first jumper wire to the left side of resistor R_2. Upper terminal strip 14 electrically connects the right side of resistor R_2 to the top side of the second jumper wire, which bridges the isolating trench between the upper and lower terminal strips. Lower terminal strip 14 electrically connects the bottom side of the second jumper wire to the right side of resistor R_3. Lower terminal strip 9 electrically connects the left side of resistor R_3 to the top side of the third jumper wire. The lower bus strip Y electrically connects the bottom side of the third jumper wire to the wire to ground, completing the circuit wiring.

You can probably see that a description of the wiring on a protoboard for even a simple circuit can be quite lengthy and tedious to read, write, and follow. Technicians and engineers typically employ a *wiring list* that uses the designators on the protoboard to describe and verify the connections for a circuit. The following table is an example of a wiring list for the above circuit.

Component	Reference	Connect 1	Connect 2	Connect 3	Verified
Power	+5V	Supply	X1	NA	
1.0 kΩ	R1	X3	D3	NA	
Jumper	J1	E3	E9	NA	
2.2 kΩ	R2	C9	C14	NA	
Jumper	J2	E14	F14	NA	
56 kΩ	R3	H14	H9	NA	
Jumper	J3	I14	Y14	NA	
Ground	GND	Y1	Supply	NA	

The wiring list provides a "Verified" column that allows the assembler, technican, or engineer to record (with a check mark, the date and time, or his or her initials) that the connection has been made. This is useful not only when verifying that the circuit is wired correctly, but also to keep track of the progress in constructing the circuit should the circuit require more than one session to complete. Note that the wiring list also provides columns for three connections to accommodate three-terminal devices like transistors, even though the example circuit contains only two-terminal devices. For a three-lead component, the circuit assembler should note the specific terminal (such as E, B, C) as well as the protoboard location for each of its connections. For devices with four or more terminals, the wiring list must be modified accordingly.

Some Tips on Protoboarding Circuits

Protoboards are very useful for constructing fast prototypes, but suffer from some limitations. One obvious limitation is that surface-mount devices require special preparation (such as wires soldered to their leads or special leaded carriers) for use on protoboards. Another limitation is that the spring clips and jumper wires limit protoboard circuit frequencies to several megahertz. Below are some other tips for using protoboards.

- Always develop a wiring diagram for a circuit. A simple way to do so is to create blank wiring lists and fill them in as you build or verify the circuit.

- Use only solid wire for jumpers. Untinned stranded wires can break, leaving conductive "whiskers" on and inside the board that could short out components. Tinned stranded wires can damage the spring clips.

- Never attempt to insert wires or leads that are too large (over AWG20) for the holes in a protoboard. This will stress or damage the spring clips inside the protoboard so that they may no longer properly grip smaller gauge leads and cause erratic operation in future circuits.

- Do not exceed the recommended bend radius of component leads. This can fracture the lead or mechanically damage the internal component connections. If possible, use a bending jig when forming the leads to prevent mechanically stressing the part.

- Trim component and jumper leads so that the body does not quite touch the surface of the protoboard when the leads are fully inserted. This will minimize the risk of someone touching or shorting out component leads or snagging and pulling out components from the board.

- Minimize the distance for connections between components while leaving some space between physically adjacent components. This makes it easier to trace component connections and modify circuits. It also ensures that components have adequate clearance so that they can effectively dissipate power.

- Develop the habit of color-coding circuit wiring, especially for power, ground, and digital bus connections. A common practice is to use red for +5 V, black for ground, and the resistor color code values for bus lines (e.g., black for D_0, brown for D_1, red for D_2, etc.). This will greatly simplify troubleshooting circuits with many connectors.

- Be consistent when assigning voltages to bus strips. This will help to avoid many problems with damaging components.

Acknowledgments

The authors would like to express their appreciation to all those who helped with the development and production of this manual. In particular, they would like to thank Rex Davidson, Wyatt Morris, and other Pearson Education staff for all their hard work and assistance, Lois Porter for her thorough technical editing work, and Mark Walters and Tien Pham of National Instruments for their technical assistance with Multisim.

DC/AC Experiments

DC/AC Experiment 1 - Introduction to Multisim for DC/AC Circuits.............3

DC/AC Experiment 2 - Basic Electronic Concepts................................9

DC/AC Experiment 3 - Ohm's Law and Watt's Law 27

DC/AC Experiment 4 - Series Resistive Circuits................................35

DC/AC Experiment 5 - Parallel Resistive Circuits.............................39

DC/AC Experiment 6 - Series-Parallel Resistive Circuits 45

DC/AC Experiment 7 - Relays...51

DC/AC Experiment 8 - AC Measurements 57

DC/AC Experiment 9 - Capacitors 65

DC/AC Experiment 10 - AC Response of *RC* Circuits......................71

DC/AC Experiment 11 - Inductors 79

DC/AC Experiment 12 - AC Response of *RL* Circuits......................85

DC/AC Experiment 13 - Series and Parallel Resonance....................95

DC/AC Experiment 14 - Transformers 101

DC/AC Experiment 15 - Time Response of Reactive Circuits 107

Name _____   Class _____

Date _____   Instructor _____

DC/AC Experiment 1 - Introduction to Multisim for DC/AC Circuits

1.1 Introduction

The world of electronics is continually changing. To remain competitive, companies with electronic products must rapidly design, prototype, test, manufacture, and market their products. Two-sided boards and leaded components were once the industry standard, and companies could use perfboard, protoboards, and wire-wrap boards to evaluate new parts and to build and test circuit prototypes. In modern electronics multilayer boards and surface-mount components are the standard, so creating physical prototypes from scratch is both problematic and expensive. To help solve this problem, component manufacturers often provide evaluation kits so that customers can determine whether a new part will meet their needs and better understand the practical requirements for using the part in circuits. Another valuable resource is specialized software to quickly design, evaluate, and modify circuit designs.

The Multisim software is a program that acts as a virtual electronics laboratory. You can use the Multisim program not only to create electronic circuits on your computer, but also to simulate ("run") the circuits and use virtual laboratory instruments to make electronic measurements. Simulation software, such as Multisim, has become an integral part of the design process for new electronic circuits.

In Part 1 of this experiment, you will familiarize yourself with the Multisim interface, particularly those components associated with basic dc and ac electronics. In Part 2, you will explore Multisim by identifying various Multisim tools and their associated toolbars.

1.2 Reading

National Instruments, *NI Circuit Design Suite: Getting Started with Circuit Design Suite*, Chapters 1 and 2

1.3 Key Objectives

Part 1:   Start the Multisim program and learn the components of the Multisim user interface.

Part 2:   Identify selected Multisim toolbars and tools.

Part 1: Working with Multisim Software

1.4 Starting the Multisim Program

1.4.1 Starting the Multisim Program from the Windows Desktop

If you added a Multisim shortcut to your desktop during installation, then you can use the desktop shortcut to start the Multisim program. To do so:

1)   Navigate to the desktop.

2)   Double-click the Multisim icon. Your icon may differ from that of Figure 1-1.

Figure 1-1: Multisim Icon

1.4.2 Starting the Multisim Program from the Start Menu

Note that the following instructions assume that you accepted the default locations for your version of the Multisim software.

1)   Click the **Start** button in the lower left corner of the Windows screen.

2)   Click **Programs**.

3) Click the "National Instruments" folder.

4) Click the "Circuit Design Suite" folder for your version of Multisim.

5) Click the appropriate file for your version of Multisim.

1.4.3 Starting the Multisim Program Using the Run… Command

Note that the following instructions assume that you accepted the default locations for your version of the Multisim software.

1) Click the **Start** button in the lower left corner of the Windows screen.

2) Click **Run…**.

3) Click the **Browse…** button.

4) Select the "C:" drive in the **Look in:** drop-down list.

5) Double-click the "Program Files" folder.

6) Double-click the appropriate "National Instruments" folder.

7) Double-click the appropriate "Circuit Design Suite" folder for your version of Multisim.

8) Double-click the appropriate file for your version of Multisim.

Alternatively, you can enter the path to the Multisim executable file in the **Open…** text box and click the **OK** button.

You may wonder why you would ever use the **Run…** command to start Multisim or any other Windows application. The answer is that once you have used this method the Windows operating system will remember the path to the Multisim program. You can then select Multisim directly from the drop-down list in the **Run…** window, unless your computer's security settings delete your session history.

1.5 The Multisim Interface

Once the Multisim program starts, you will see the screen (or one much like it) shown in Figure 1-2.

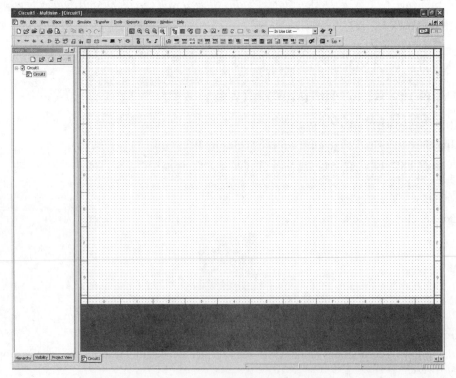

Figure 1-2: Multisim Interface

In addition to the standard Windows title bar, menu bar, toolbar, and screen controls, the Multisim interface contains a number of special toolbars that allow you to create and simulate circuits. The number and type of toolbars that you will see in the Multisim interface depends upon which toolbars you enable in the **View →** **Toolbars...** menu. The following sections provide a brief description of some of the toolbars that you will use in this workbook. The figures show toolbars from the Multisim software.

1.5.1   The Title Bar

The **Title** bar is the region at the very top of the screen and is common to all Windows applications. The left side of the bar contains information about the application and file, while the right side of the bar contains controls to minimize, maximize, and close the application.

1.5.2   The Menu Bar

The **Menu** bar, shown in Figure 1-3, contains standard Windows menus (such as **File**, **Edit**, **View**, and **Help**) and menus that are specific to Multisim (such as **Place**, **MCU**, **Simulate**, and **Reports**). These menus let you configure and operate the application.

Figure 1-3: The Multisim Menu Bar

1.5.3   The Standard Toolbar

The **Standard Toolbar**, shown in Figure 1-4, contains tools for performing common operations from the File and Edit menus, such as creating a new file, printing a file, and pasting information from the Clipboard into the open file.

Figure 1-4: The Standard Toolbar

1.5.4   The View Toolbar

The **View Toolbar**, shown in Figure 1-5, contains tools for performing common Windows operations from the View menu, such as zooming in on or zooming out from the current view.

Figure 1-5: The View Toolbar

1.5.5   The Main Toolbar

The **Main Toolbar**, shown in Figure 1-6, contains tools to access various Multisim features and information about the current circuit.

Figure 1-6: The Main Toolbar

1.5.6   The Simulation Run Toolbar

The **Simulation Run Toolbar**, shown in Figure 1-7, contains tools that let you start, stop, pause, and resume a simulation. Note that the **Pause** control will be grayed out unless you have started a simulation.

Figure 1-7: The Simulation Run Toolbar

1.5.7   The Component Toolbar

The **Component Toolbar**, shown in Figure 1-8, contains tools that let you access various components with which to create and analyze circuits.

Figure 1-8: The Component Toolbar

The component tools with which you will most often work for basic dc and ac electronics are the **Place Source** and **Place Basic** tools, shown in Figure 1-9.

Figure 1-9: "Place Source" and "Place Basic" Tools

1.5.8   The Instruments Toolbar

The **Instruments Toolbar**, shown in Figure 1-10, provides instruments with which you can measure and evaluate the operation of circuits. Some instruments, like the multimeter, oscilloscope, and logic analyzer, are real devices that technicians and engineers use to analyze real-world circuits. Other instruments, like the Bode plotter and logic converter, exist only within the Multisim application and are convenient tools for you to simulate, analyze, and debug your circuit designs.

Figure 1-10: The Instruments Toolbar

The instrument tools with which you will most often work for basic dc and ac electronics are the **Multimeter**, **Function Generator**, and the **2-Channel Oscilloscope** and **4-Channel Oscilloscope** tools, shown in Figure 1-11.

Figure 1-11: Multimeter, Function Generator, 2-Channel Oscilloscope, and 4-Channel Oscilloscope Tools

You may have noticed that the two tools on the far right of the **Instruments Toolbar** have small arrows pointing down next to them. These are "flyouts" that open menus that provide further selections for the tools.

1.5.9   Tool Tips

With so many toolbars and tools available, the Multisim interface may seem rather confusing at first. To help you, the Multisim software uses tool tips to help you identify each tool. A tool tip is a small text box that tells you what the tool is. To activate a tool tip, simply let the pointer hover, or remain over, the tool you wish to identify. After a short time the tool tip for that tool will appear, as shown in Figure 1-12. The tool tip indicates that the highlighted instrument (the tool with the border around it) is the function generator.

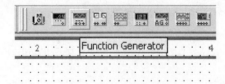

Figure 1-12: Function Generator Tool Tip

Note that the **Pause** control does not have a tool tip associated with it.

In the exercise in Part 2 you will use the tool tips to identify a number of tools and better familiarize yourself with some of the Multisim tools.

1.6 Closing the Multisim Circuit

There are two standard ways to close the Multisim circuit. If you made any changes to the circuit file (even if the net effect of those changes did not change the circuit file, such as adding a component and then deleting it), then the program will display a dialog box similar to that in Figure 1-13.

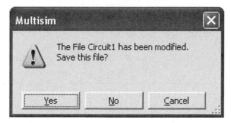

Figure 1-13: File Save Reminder Dialog Box

If you receive this notification, click the **No** button. You will learn about saving files in the next section.

1.6.1 Closing the Multisim Circuit from the File Menu

To close the Multisim circuit from the **File** menu:

1) Click **File** in the Multisim **Menu** bar.
2) Click **Close**.

1.6.2 Closing the Multisim Program with the Application Close Button

To close the Multisim program with the **Application Close** button, click the ☒ (**Close**) button at the far right of the blue title bar. If you click the **Close** button in the Multisim **Menu** bar, then you will close the circuit file but not the program.

Part 2: Locating Tools in Multisim

Now that you have some experience with Multisim, this exercise will help you to better familiarize yourself with the user interface.

1) Use one of the methods in Section 1.4 to start the Multisim program.
2) Use the information in Section 1.5 and the Multisim tool tips to complete Table 1-1. Indicate
   - the toolbar on which the tool appears, and
   - the identity of the tool.

Table 1-1: Multisim Tool Identification (continued on next page)

Tool Graphic	Associated Toolbar	Tool Identity
⊥ₖ		
▦		
▦		

Tool Graphic	Associated Toolbar	Tool Identity

DC/AC Experiment 2 - Basic Electronic Concepts

2.1 Introduction

You can describe the operation of all electronic circuits in terms of basic electrical quantities. Some electrical quantities, such as voltage, are probably already familiar to you. Other quantities, such as reactance, may be new. Electronic components provide a practical means for you to work with electrical quantities so that you can design and construct a circuit. A battery, for example, can supply the voltage to a circuit, and resistors can determine how much current will flow in specific parts of the circuit. Once you understand electrical quantities you can understand the role of specific components and the operation of circuits containing them.

In addition to understanding what electrical quantities are, you must also have some way to measure these quantities. Designing a 12-volt power supply is pointless (and probably impossible) if you cannot determine how much a "volt" is. You must be able to measure electrical quantities like voltage and current to also determine whether or not a circuit is functioning correctly, and to isolate any problems in the circuit.

In Part 1 of this experiment, you will review basic electrical quantities, the units for each, and the components that are associated with electrical quantities. In Part 2, you will learn measurement techniques for voltage, current, and resistance.

2.2 Reading

Floyd and Buchla, *Electric Circuits Fundamentals, 8th Ed.*, Chapters 1 and 2.

National Instruments, *NI Circuit Design Suite: Getting Started with Circuit Design Suite*, Chapter 2

2.3 Key Objectives

Part 1:   Review the basic concepts of voltage, current, and resistance and their associated units. Place and connect voltage source, current sources, and resistors in the Multisim program.

Part 2:   Use the Multisim digital multimeter (DMM) to make voltage, current, and resistance measurements.

2.4 Multisim Files

Part 2:   *DC_AC_Exp_02_Part_02a, DC_AC_Exp_02_Part_02b*, and *DC_AC_Exp_02_Part_02c*

Part 1:   Voltage, Current, and Resistance

2.5 Voltage

Voltage, typically represented by *V* and occasionally by *E*, is measured in **volts** (V). Separating particles with opposite charges requires work, and this separation increases the potential energy of the charges. Voltage represents the potential energy of the separated charges and is equal to their difference in potential energy *W* per charge *Q* so that $V = W / Q$. A common analogy for voltage is that of water pressure in a gravity-fed water system. Energy in the form of work raises the water in the system to a storage tank. The water that is stored in the tank exerts pressure and possesses potential energy that can perform work when released. Think of voltage as the electrical "pressure" across a circuit.

Voltage sources apply a specific amount of voltage across a circuit. There are two fundamental types of voltage sources. One is the ac voltage source, in which the polarity of the voltage periodically reverses. The other is the dc voltage source, in which the polarity of the voltage does not change. Figure 2-1 shows the symbols for ac and dc voltage sources.

V1
120 Vrms
60 Hz
0°

V2
12 V

AC VOLTAGE SOURCE   DC VOLTAGE SOURCE

Figure 2-1: AC and DC Voltage Source Symbols

2.6 Current

Current, typically represented by I, is measured in **amperes** (A), commonly shortened to "amps." Electrical current is the net movement of charge. Just as you can think of voltage as the electrical pressure in a circuit, current is the electrical movement of charge that results.

Although voltage sources are more commonly used in electronic circuits, you will sometimes encounter current sources. Current sources, similar to voltage sources, inject a specific amount of current into a circuit rather than apply a specific amount of voltage across it. Figure 2-2 shows the symbols for ac and dc current sources.

I1
1 A
1kHz
0°

I2
1 A

AC CURRENT SOURCE   DC CURRENT SOURCE

Figure 2-2: AC and DC Current Source Symbols

2.7 Resistance

Resistance, typically represented by R, is measured in **ohms** (Ω). Resistance is the opposition to current. Just as the diameter of a pipe determines how much water will flow for a given amount of water pressure, resistance determines how much current will flow for a given voltage.

The **resistor** provides a known amount of resistance in a circuit. Resistors can be either fixed or variable, as shown in Figure 2-3. The symbol on the left shows a fixed resistor, which is a two-terminal device whose value cannot change. The symbol on the right shows a three-terminal device called a potentiometer, sometimes mistakenly referred to as a variable resistor. The resistance value between the top and bottom terminals is fixed. The resistance between the third terminal (called the wiper) and the top or bottom terminal depends upon its physical position, which the user can vary. As the wiper moves towards the top terminal, the resistance between the top terminal and wiper decreases and the resistance between the bottom terminal and wiper increases. Conversely, as the wiper moves towards the bottom terminal, the resistance between the top terminal and wiper increases and the resistance between the bottom terminal and wiper decreases.

R1
1.0kΩ

R2
1.0kΩ   50%
Key=A

FIXED RESISTOR   POTENTIOMETER

Figure 2-3: Resistor Symbols

If you connect the wiper to one of the other terminals, as shown in Figure 2-4 , you will create a two-terminal variable resistor. The resistance between the top and bottom terminals will change as the position of the wiper changes.

VARIABLE RESISTOR

Figure 2-4: Potentiometer Connected as a Variable Resistor

2.8   Building a Simple Multisim Circuit

2.8.1   Placing Components

For this part of the experiment, you will use the Multisim program to place and connect components.

1)   Use one of the methods from Section 1.4 to start the Multisim program.

2)   Click the **Place Source** tool in the **Component Toolbar** to open the component browser in Figure 2-5.

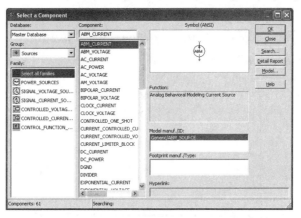

Figure 2-5: Sources Component Browser

3)   Select "POWER_SOURCES" from the **Family:** window and "DC_POWER" from the **Component:** list. Note that the symbol in the window changes to that of a battery (i.e., a dc source).

4)   Click the **OK** button. You will return to the circuit window in the Multisim interface. Note that as you move the pointer, the component symbol moves with it. Move the battery symbol to a position near the left center of the circuit window (refer to Figure 2-6) and click to place it.

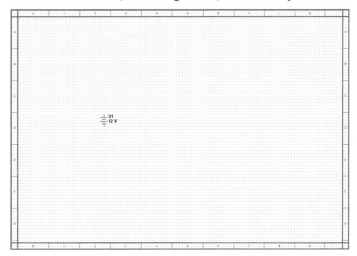

Figure 2-6: Circuit Window with Battery Placed

At this point, depending on your global preferences, the program may automatically return you to the component browser. If it does not, click the **Place Source** tool again.

5) In the component browser, select "GROUND" from the **Component:** list.

6) Click the **OK** button and place the ground symbol beneath the battery symbol, as shown in Figure 2-7.

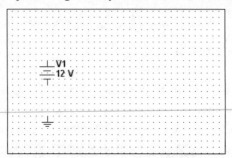

Figure 2-7: Circuit with Ground Placed

7) Place a second ground symbol to the right of the other symbols on the board, as shown in Figure 2-8. This completes the power source symbols you will need to construct the initial circuit.

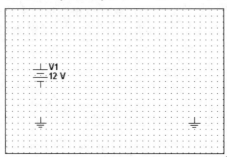

Figure 2-8: Circuit with Second Ground Placed

8) Click the **Place Basic** tool in the **Component Toolbar**. The component browser appears with the basic components, as shown in Figure 2-9.

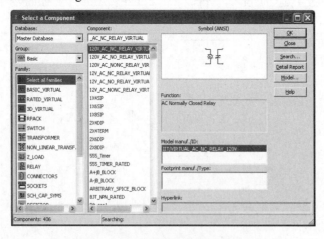

Figure 2-9: Basic Component Browser

9) Select "RESISTOR" from the **Family:** window and "1k" from the **Component:** list. If you do not see "RESISTOR" in the **Family:** list or "1k" in the **Component:** list, use the vertical sliders on the right side of the lists to scroll through the list until you see them.

10) Click the **OK** button and place the resistor symbol above the right ground, as shown in Figure 2-10.

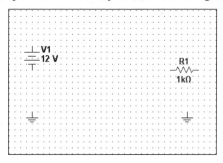

Figure 2-10: Circuit with Initial Resistor Placement

As you can see, the resistor orientation is horizontal. This will not affect the circuit but the resistor would look better if its orientation was vertical.

11) Right-click on the resistor. This will select R_1, as indicated by a dashed box around the resistor, and open the component right-click menu, shown in Figure 2-11.

Figure 2-11: Component Right-Click Menu

12) Click on **90 Clockwise** to rotate the resistor 90 degrees in the clockwise direction, as shown in Figure 2-12. Alternatively, you can press "w" on the keyboard when the right-click menu is open, or "CTRL + R" (hold the "CTRL" key down while pressing the "R" key) when the resistor is selected to rotate the resistor.

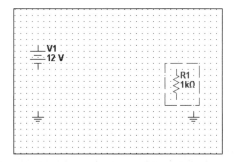

Figure 2-12: Circuit with R_1 Rotated

13) With the R_1 still selected, position the cursor over the resistor. Left-click on the resistor and hold the mouse button down. While holding the mouse button down, move the mouse to drag the resistor so that it lines up with the right ground, as shown in Figure 2-13.

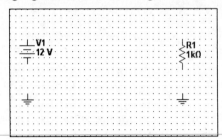

Figure 2-13: Circuit with R_1 Properly Positioned

2.8.2 Connecting Components

The Multisim program has no **Wire Tool** to connect components. You create a wire by clicking in the workspace to select the start and end points for the wire.

Note that the program will not let you create a wire that does not begin on a component terminal or end on a wire segment or component terminal. If you wish to begin a wire segment on a wire segment or leave one end of a wire segment unconnected you must connect the wire to a special terminal called a junction. To place a junction, select **Junction** from the **Place** menu. When you end a wire on an existing wire segment, the Multisim program automatically creates a junction for you.

1) Move the tip of the pointer near the cathode (negative terminal) at the bottom of the battery. The pointer will change from an arrow to a small crosshair, indicating that it is ready to make a connection to the battery terminal.

2) Click the left mouse button and drag the crosshair towards the left ground symbol. As you do so, you will see a wire being drawn. If you do not see it, position the pointer near the battery and left-click again when the crosshair is visible.

3) When you reach the terminal at the top of the ground symbol, left-click the mouse. This will connect the wire from the battery to the ground symbol, as shown in Figure 2-14. Depending on your program settings, you may or may not see the net name (in this case "0") for the wire. To show or hide net names:

   a. Select **Sheet Properties...** from the **Options** menu in the menu bar.

   b. In the **Net Names** field of the **Circuit** tab, choose "Show All" to show the names of all nets and "Hide All" to hide the names of all nets. "Use Net-specific Setting" will show only the nets that you individually configure to show the net name. This example used the "Hide All" setting.

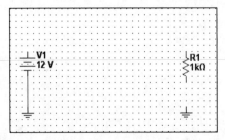

Figure 2-14: Circuit with Cathode of V_1 Connected to Ground

4) Finish connecting the remaining components, as shown in Figure 2-15.

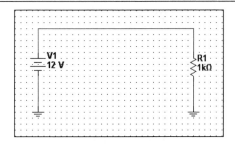

Figure 2-15: Completed Circuit

Congratulations! You've just completed your first Multisim circuit!

2.9   Saving a Multisim Circuit File

A good idea is always to save your work, especially after working for more than a few minutes. There are two methods for saving a Multisim circuit file. If you wish, use one of the standard methods below to save your circuit to your computer. A good rule is to always use a filename that accurately describes the circuit, such as *Experiment_02_12V_Voltage_Source*.

2.9.1   Using the Save File Tool to Save a Circuit File

1) Click the **Save File** tool on the **Standard Toolbar**.

2) If you have saved the file before, the Multisim program will use the same name and location as before to save the file. If you are saving the circuit for the first time the **Save As** window will appear, as shown in Figure 2-16.

Figure 2-16: Multisim Save As Window

3) Use the **Save in:** drop-down list to navigate to the folder in which you wish to save the file, and enter the file name you wish to use in the **File name:** box.

4) Once you have finalized the file name and location for your circuit file, click the **Save** button.

5) Select **Close** from the **File** menu to close the circuit.

2.9.2   Using the File Menu to Save a Circuit File

1) Click **File** on the **Menu** bar.

2) If you have previously saved the file and wish to save the file under the same name, click **Save** on the **File** menu. Otherwise, click **Save As…** on the **File** menu. The **Save As** window will appear, as shown in Figure 2-16.

3) Use the **Save in:** drop-down list to navigate to the folder in which you wish to save the file, and enter the name of the file you wish to use in the **File name:** box.

4) Once you have finalized the file name and location for your circuit file, click the **Save** button.

5) Select **Close** from the **File** menu to close the circuit.

2.10 Opening an Existing Multisim Circuit

You can use one of the standard methods below to open a Multisim circuit that you or someone else created and saved. The Multisim program can open circuit files created by earlier versions of Multisim.

2.10.1 Using the Open File Tool to Open a Circuit File

1) Click the **Open File** tool on the **Standard Toolbar**. The **Open File** window will appear, as shown in Figure 2-17.

Figure 2-17: Multisim Open File Window

2) To specify the file you wish to open, you can
   - use the **Look in:** and **Files of type:** drop-down lists to locate the file,
   - use the **File name:** drop-down list to select the file from files you have previously opened, or
   - enter the path and name of the file in the **File name:** box.
3) Once you have selected a file, click the **Open** button.

2.10.2 Using the File Menu to Open a Circuit File

1) Click **File** on the **Menu** bar.
2) Click **Open...** on the **File** menu. The **Open File** window will appear, as shown in Figure 2-17.
3) To specify the file you wish to open, you can
   - use the **Look in:** and **Files of type:** drop-down lists to locate the file,
   - use the **File name:** drop-down list to select the file from files you have previously opened, or
   - enter the path and name of the file in the **File name:** box.
4) Once you have selected a file, click the **Open** button.

2.10.3 Using the File Menu to Open a Recent Circuit Design

If you wish to open a circuit on which you have recently worked, you can select it from a list of recent files.

1) Click **File** on the **Menu** bar.
2) Move the cursor over **Recent Designs** ▶ on the **File** menu. A list of your recent Multisim circuit files will appear.
3) Click on the circuit file you wish to open.

2.10.4 Using Windows Explorer to Open a Circuit File

1) If the Multisim program is not running, you can use Windows Explorer to both start the Multisim program and open the circuit file.
2) Open Windows Explorer.

3) Navigate to the folder that contains the circuit file you wish to open.

4) Double-click the circuit file. Windows will start the Multisim program, and the Multisim program will load the circuit file.

2.11 Modifying an Existing Circuit

Design is an ongoing process. At some point you will probably wish to change a circuit, whether to correct an error, move to the next phase of an incremental design process, improve a design, or just to tinker and see what the change will do. Changes to designs in the past were laborious and time-consuming. The Multisim program offers a number of features and tools with which you can quickly and easily change existing circuits. In this section you will learn some common ways to modify an existing circuit in the Multisim environment.

2.11.1 Changing Component Values

A common design modification is changing the value of a component. For this part of the experiment, you will change the value of the voltage source from 12 volts to 6 volts and the value of the resistor from 1 kΩ to 510 Ω. First, change the value of the voltage source.

1) Use one of the methods in Section 2.10 to load your circuit.

2) Right-click on the voltage source to open the component right-click menu.

3) Click on **Properties** to open the **DC_POWER** properties window, as shown in Figure 2-18.

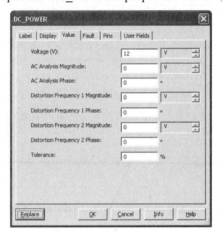

Figure 2-18: DC_POWER Properties Window

4) Click in the **Voltage (V):** window and drag the cursor to select and highlight the current value "12".

5) Type the new value "6" in the **Voltage (V):** window.

6) Click the **OK** button to make the change. Your circuit should now look like Figure 2-19.

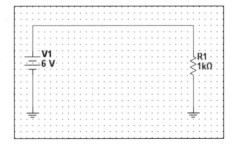

Figure 2-19: Circuit with Modified Voltage Source

Now, modify the resistor.

7) Right-click on the resistor to open the component right-click menu.

8) Click on **Properties** to open the **RESISTOR** properties window, as shown in Figure 2-20.

Figure 2-20: DC_POWER Properties Window

9) Click in the **Resistance (R):** window to select and highlight the current value "1k".

10) Type the new value "510" in the **Resistance (R):** window.

11) Click the **OK** button to make the change. Your circuit should now look like Figure 2-21.

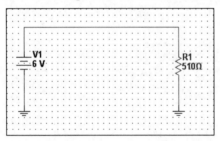

Figure 2-21: Circuit with Modified Resistance Value

12) Use **Save As...** in the **File** menu to save your modified file with a new filename.

13) Close the circuit file.

2.11.2 Replacing Components

Another common circuit modification is to replace an existing component with another type of component. For this part of the experiment, you will change the 12 V dc voltage source in your original circuit to a 10 mA dc current source. First, replace the voltage source with a current source.

1) Use one of the methods in Section 2.10 to load your original circuit.

2) Right-click on the voltage source to open the component right-click menu.

3) Click on **Properties** to open the **DC_POWER** properties window, as shown in Figure 2-22.

Figure 2-22: DC_POWER Properties Window

4) Click the **Replace** button to open the **SELECT A COMPONENT** window, as shown in Figure 2-23.

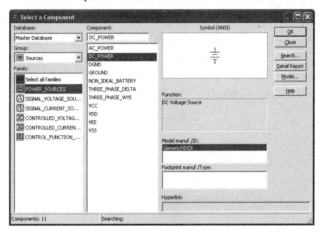

Figure 2-23: SELECT A COMPONENT Window

5) Select "SIGNAL_CURRENT_SOURCES" from the **Family:** list.

6) Select "DC_CURRENT" from the **Component:** list.

7) Click the **OK** button to make the change. Your circuit should now look like Figure 2-24.

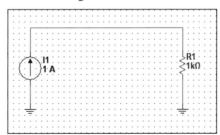

Figure 2-24: Circuit with DC Current Source

Now, change the value of the current source from 1 A to 10 mA.

8) Right-click on the current source to open the component right-click menu.

9) Click on **Properties** to open the **DC_CURRENT** properties window, as shown in Figure 2-25.

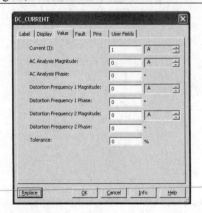

Figure 2-25: DC_POWER Properties Window

10) Click in the **Current (I):** window and drag the cursor to select and highlight the current value "1".

11) Type the new value "10" in the **Current (I):** box.

12) Click the down arrow (▼) for the current units box once to change the units from "A" to "mA".

13) Click the **OK** button to make the change. Your circuit should now look like Figure 2-26.

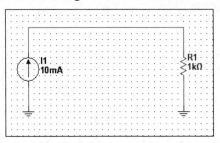

Figure 2-26: Circuit with Modified DC Current Source

14) Use **Save As...** in the **File** menu to save your modified file with a new filename.

15) Close the circuit file.

Questions for Part 1

1) What is the main difference between a dc and an ac source?

2) What is the difference between a voltage source and a current source?

3) You are prototyping a circuit and are not sure of the exact value of resistance the circuit will require. Would you probably use a fixed or a variable resistor to prototype the circuit? Why?

Part 2: Multisim Circuit Measurements

The most common circuit measurement tool is the multimeter. Basic multimeters combine the functions of a voltmeter, ammeter, and ohmmeter to measure voltage, current, and resistance, respectively. More expensive multimeters can measure capacitance, test diodes, and perform other useful functions. The Multisim program provides a **Multimeter** tool on the **Instruments Toolbar**, which is circled in Figure 2-27.

Figure 2-27: The Multimeter Tool on the Instruments Toolbar

Figure 2-28 shows the multimeter in both minimized and expanded form. Table 2-1 summarizes the multimeter functions.

Figure 2-28: Multisim Multimeter Views

Table 2-1: Multimeter Functions

Button	Function
A	Selects ammeter mode.
V	Selects voltmeter mode.
Ω	Selects ohmmeter mode.
dB	Selects decibel mode.
∿	Selects ac measurement mode.
—	Selects dc measurement mode.
Set...	Opens multimeter settings dialog box.

2.12 Measuring Voltage

In this exercise you will use the Multisim multimeter to measure the voltages in a circuit. All voltage measurements are with respect to some reference point. Usually this reference point is circuit ground (which is considered to represent 0 V) but need not be. A common convention is to specify a voltage at point X with respect to ground as V_X, and the voltage at point X with respect to point Y as V_{XY}.

Voltage across a component is often represented as V_{REFDES}, where "REFDES" is the reference designator for the component. In Figure 2-29 XMM_1 is measuring V_A, the voltage at point A with respect to ground, XMM_3 is measuring V_B, which is the voltage at point B with respect to ground, and XMM_2 is measuring V_{AB}, which is the voltage at point A with respect to the voltage at point B, or $V_A - V_B$. This voltage is also referred to as V_{R1}, as this is the voltage across R_1.

Voltage has polarity. Connecting the positive (+) lead of an analog meter to the negative (−) terminal of a dc source and vice versa could damage the movement of an analog meter. This is not the case with digital meters. If the voltage on its positive terminal is more positive than the voltage on its negative terminal, a DMM will indicate that the voltage is positive. If the voltage on its positive terminal is more negative than the voltage on its negative terminal, the DMM will indicate that the voltage is negative.

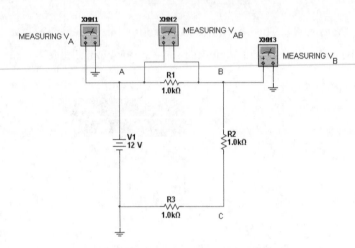

Figure 2-29: Voltage Measurements

1) Open the Multisim file *DC_AC_Exp_02_Part_02a*.

2) Select the Multimeter tool from the **Instruments Toolbar**.

3) Select the dc measurement mode.

4) Select the voltmeter mode.

5) Connect the negative (−) terminal of the multimeter to ground.

6) Connect the positive (+) terminal of the multimeter to point A of the circuit.

7) Start the simulation. The multimeter will work only when Multisim is simulating a circuit.

8) Record the voltage for V_A in Table 2-2.

9) Stop the simulation.

10) Repeat Steps 5 through 9 for points B through D to measure and record each of the voltages V_B through V_D for the circuit.

Table 2-2: Measured Voltage Values Relative to Ground

Voltage	Value
V_A	
V_B	
V_C	
V_D	

11) Use the measured voltages in Table 2-2 to calculate and record the values for the "Calculated" column in Table 2-3.

12) Connect the negative (−) terminal of the multimeter to point B.

13) Connect the positive (+) terminal of the multimeter to point A.

14) Start the simulation.

15) Record the voltage for V_{R1} (or $V_A - V_B$) in the "Measured" Column in Table 2-3.

16) Stop the simulation.

17) Repeat Steps 12 through 16 to measure and record the voltages for V_{R2} through V_{R4} in Table 2-3.

Table 2-3: Voltage Values Across Resistors

Voltage Difference	Calculated	Resistor Voltage	Measured
$V_A - V_B$		V_{R1}	
$V_B - V_C$		V_{R2}	
$V_C - V_D$		V_{R3}	
V_D – Ground (0 V)		V_{R4}	

Observations:

2.13 Measuring Current

In this exercise you will use the Multisim multimeter to measure the currents in a circuit. The Multisim multimeter differs from practical multimeters in that practical meters have different lead connections for voltage and current measurements. This is because, unlike voltage measurements, you must insert the multimeter into the circuit so that current passes directly through the meter and the meter can measure the current. *NEVER* connect an ammeter across a component to measure current, as very high current could flow through and damage the meter.

Sometimes it is not practical to measure current directly, such as when you wish to measure a varying or alternating current. In these cases the circuit may contain a sense resistor. A sense resistor is a small-valued resistor through which the current you wish to measure passes. You can measure the voltage across the resistor and use the value of the resistor to calculate the current value.

Refer to Figure 2-30 for the correct and incorrect ways to connect an ammeter to a circuit.

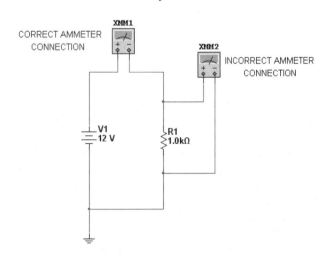

Figure 2-30: Correct and Incorrect Ammeter Connections for Current Measurements

Current, like voltage, has polarity. Connecting the leads of an analog meter incorrectly so that current flows in the wrong direction through the meter movement could damage the meter. This is not the case with digital meters. A DMM will indicate conventional current entering the positive (+) terminal as positive and current entering the negative (−) terminal as negative.

A common convention is to identify the current through a component as I_{REFDES}, where "REFDES" is the reference designator for the component. In Figure 2-30 the current through R_1 would be I_{R1}.

1) Open the Multisim file *DC_AC_Exp_02_Part_02b*.

2) Select the Multimeter tool from the **Instruments Toolbar**.

3) Select the dc measurement mode.

4) Select the ammeter mode.

5) Delete the wire between V_S and R_1.

6) Connect the negative (+) terminal of the multimeter to the positive terminal of V_S.

7) Connect the positive (−) terminal of the multimeter to point A of the circuit.

8) Start the simulation. The multimeter will work only when Multisim is simulating a circuit.

9) Record the current I_{R1} to three significant digits for $V_S = 1.0$ V in Table 2-4.

10) Stop the simulation.

11) Change the value of V_S to the next value shown in Table 2-4.

12) Repeat Steps 8 through 11 for each value of V_S in Table 2-4.

Table 2-4: Measured Current Values

V_S	I_{RI}	V_S	I_{RI}
1 V		50 V	
2 V		100 V	
5 V		200 V	
10 V		500 V	
20 V		1000 V	

Observations:

2.14 Measuring Resistance

In this exercise you first will use the resistor color codes to determine the nominal values for the resistors in a circuit, and then use the multimeter to measure the actual resistor values and determine whether there is a fault with the resistor. Although Multisim can specify resistor tolerance (how much the actual resistance differs from the nominal, or ideal, resistance), this exercise omits the resistor tolerance for simplicity.

An ohmmeter measures current by applying a fixed voltage across a component and measuring the resulting current. You must remove power from the circuit when making resistance measurements, and you will typically disconnect the component from the circuit as well. You connect the meter across the points whose resistance you wish to measure. As with voltage measurements, a common convention is to identify the measured resistance as R_{XY}, where X and Y are the points to which you connect the meter. The multimeter in Figure 2-31 is measuring R_{DE}. As you will learn in later experiments, R_{DE} is not the same as R_5.

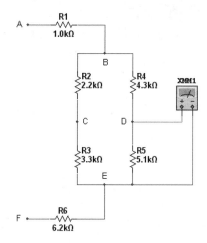

Figure 2-31: Measuring Resistance R_{DE}

1) Use the standard resistor color codes to determine and record the nominal values of R_1 through R_5 in the "Nominal" column of Table 2-5. Ignore the resistor tolerance, which is not indicated.

2) Open Multisim file *DC_AC_Exp_02_Part_02c*.

3) Select the Multimeter tool from the **Instruments Toolbar**.

4) Select the ohmmeter mode. Note that this will delete the ac mode button.

5) Delete the wire between R_1 and the rest of the circuit.

6) Connect one terminal of the multimeter to the top of R_1.

7) Connect the other terminal of the multimeter to ground.

8) Start the simulation. The multimeter will work only when Multisim is simulating a circuit.

9) Record the value of R_1 in the "Measured" column in Table 2-5.

10) Stop the simulation.

11) Repeat Steps 5 through 10 for each of the other resistors shown in Table 2-5.

Table 2-5: Nominal and Measured Resistor Values

Resistor	Band 1	Band 2	Band 3	Nominal	Measured
R_1	BROWN	ORANGE	BROWN		
R_2	BLUE	GREY	BLACK		
R_3	RED	YELLOW	BROWN		
R_4	YELLOW	ORANGE	GOLD		
R_5	WHITE	BROWN	RED		

Observations:

Questions for Part 2

1) Suppose that when you measure the voltage across R_2, the DMM reading is −1.2 V. What is the most likely explanation?

2) From the data in Table 2-2, what voltage would you expect to measure across both R_2 and R_3?

3) From the data in Table 2-4, what would you expect the approximate value of I_{R1} to be for $V_S = 15$ V?

4) From the data in Table 2-4, what would you expect the approximate value of V_S to be if $I_{R1} = 50$ mA?

5) Why do steps 6 and 7 of Section 2.14 not specify which terminal of the multimeter should connect to the top of the resistor and which terminal should connect to ground?

DC/AC Experiment 3 - Ohm's Law and Watt's Law

3.1   Introduction

One of the early investigators of electricity was Georg Simon Ohm, a German physicist teaching mathematics in Köln (Cologne), Germany, and for whom the units of resistance is named. His experiments showed that current is directly proportional to the applied voltage and inversely proportional to the circuit resistance. Based on his studies, he presented an equation that related current, voltage, and resistance. Today we know this relationship as **Ohm's law**:

$$I = V / R$$

where

I is the current in amperes, V is the voltage in volts, and R is the resistance in ohms.

In other words, if the voltage across a fixed resistance increases or decreases, the current through the resistor will increase or decrease proportionally. Ohm's law is one of the fundamental laws for analyzing electric circuits. There are two other forms of Ohm's law, which you can derive using basic algebra:

$$V = I \times R$$
$$R = V / I$$

Another fundamental law for electric circuits is **Watt's law**, named in honor of Scottish inventor James Watt. Watt's law defines power in terms of voltage and current. Specifically,

$$P = V \times I$$

where P is power in watts, V is the voltage in volts, and I is the current in amperes. Watt's law actually has two other forms, which you can derive from Ohm's law. The first, Watt's law for voltage and resistance, is

$$P = V \times I = V \times (V / R) = (V \times V) / R = V^2 / R$$

The second, Watt's law for current and resistance, is

$$P = V \times I = (I \times R) \times I = (I \times I) \times R = I^2 \times R$$

In Part 1 of this experiment, you will verify the validity of Ohm's law. In Part 2, you will investigate power in resistive circuits.

3.2   Reading

Floyd and Buchla, *Electric Circuits Fundamentals, 8th Ed.*, Chapter 3.

3.3   Key Objectives

Part 1:   Verify the three forms of Ohm's law for calculating current, voltage, and resistance.

Part 2:   Use the Multisim wattmeter to determine the power in resistive circuits and verify Watt's law.

3.4   Multisim Files

Part 1:   *DC_AC_Exp_03_Part_01*

Part 2:   *DC_AC_Exp_03_Part_02*

Part 1:   Ohm's Law

3.5   Current vs. Voltage

In this part of the experiment, you will verify that current is directly proportional to voltage and plot the response of three resistors.

1)   Use Ohm's law for current to calculate the current I_{CALC} to three significant digits for each value of V_S and R in Table 3-1.

2)   Open the file *DC_AC_Exp_03_Part_01*.

3) Connect a multimeter between V_S and R and set it for dc ammeter mode.

4) Double-click on the multimeter to expand it. Your circuit should look like that of Figure 3-1.

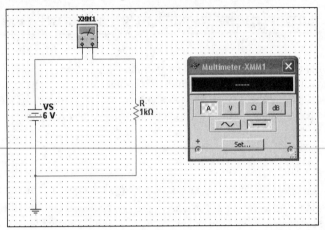

Figure 3-1: Ohm's Law Measurement Circuit 1

Figure 3-2 shows an example of the same circuit built on a protoboard.

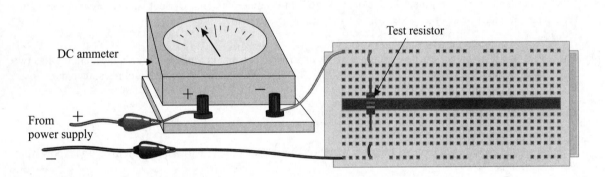

Figure 3-2: Ohm's Law Measurement Circuit on Protoboard

5) Change the value of R to 510 Ω.

6) Change the value of V_S to 1.0 V.

7) Run the simulation.

8) Record the measured current to three significant digits as I_{MEAS} for V_S and R in Table 3-1.

9) Repeat Steps 6 and 7 for the remaining values of V_S.

10) Repeat Steps 5 through 8 for $R = 1.0$ kΩ.

11) Repeat Steps 5 through 8 for $R = 2.2$ kΩ.

Table 3-1: Calculated Current Values for Fixed Resistances

V_S (V)	$R = 510\ \Omega$		$R = 1.0\ \text{k}\Omega$		$R = 2.2\ \text{k}\Omega$	
	I_{CALC} (mA)	I_{MEAS} (mA)	I_{CALC} (mA)	I_{MEAS} (mA)	I_{CALC} (mA)	I_{MEAS} (mA)
1.0						
2.0						
3.0						
4.0						
5.0						
6.0						
7.0						
8.0						
9.0						
10.0						

3.6   Resistor *V-I* Curves

In this step, you will plot current versus voltage to create what is known as a *V-I* curve. This graph shows the value of the current through a resistor for any value of voltage across it. The shape of a *V-I* curve shows how the current through a resistor changes as the voltage across it changes.

1)   For each value of V_S in Table 3-1, plot the corresponding value of I_{MEAS} for $R = 510\ \Omega$ on Plot 3-1. Label this graph "510 Ω".

2)   For each value of V_S in Table 3-1, plot the corresponding value of I_{MEAS} for $R = 1\ \text{k}\Omega$ on Plot 3-1. Label this graph "1 kΩ".

3)   For each value of V_S in Table 3-1, plot the corresponding value of I_{MEAS} for $R = 2.2\ \text{k}\Omega$ on Plot 3-1. Label this graph "2.2 kΩ".

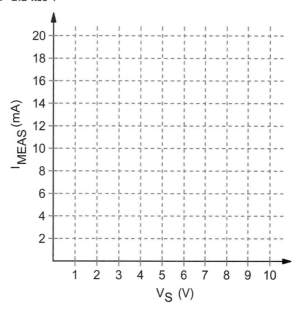

Plot 3-1: Measured Resistor *V-I* Curves

Observations:

Conclusions for Part 1

Questions for Part 1

1) One use of a *V-I* curve is to determine the approximate current through a component for a known voltage across it. From your *V-I* curve for the 510 Ω resistor, what is the approximate current when V_S = 5.5 V?

2) Another use of a *V-I* curve is to determine the approximate voltage across a component for a known current through it. From your *V-I* curve for the 2.2 kΩ resistor, what is the approximate voltage for I_{MEAS} = 3 mA?

3) The *V-I* curve for an unknown resistor lies between the *V-I* curves for the 510 Ω and 1.0 kΩ resistors. What must be true for the value of the unknown resistor?

Part 2: Watt's Law

In this part of the experiment you will learn how to use the wattmeter to measure power in a resistive circuit to examine the relationship between current, voltage, and power.

3.7 Measuring Power with the Wattmeter

The wattmeter is an actual device, although many schools and laboratories will not have one. Functionally, a standard wattmeter simultaneously measures the voltage across and the current through a device and calculates the product of the two to show the power. Figure 3-3 shows the Multisim **Wattmeter** tool in the **Instruments Toolbar**. Figure 3-4 shows the minimized and expanded views of the wattmeter.

Figure 3-3: Multisim Wattmeter Tool

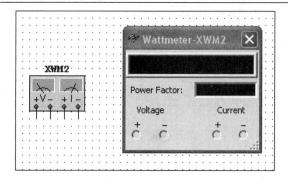

Figure 3-4: Minimized and Expanded Wattmeter Views

As Figure 3-4 shows, the connections on the left side measure the voltage, and the connections on the right side measure the current. You connect the voltage terminals as you would for a multimeter when measuring voltage, and the current terminals as you would for a multimeter when measuring current.

Note that the wattmeter displays a quantity called "Power Factor". You will learn more about this when you study ac circuits. For now, all you need to know is that the power factor in a dc resistive circuit is always 1.

1) Use Watt's law for voltage to calculate the power P_{CALC} to three significant digits for each value of V_S and R in Table 3-2.

2) Open the Multisim file *DC_AC_Exp_03_Part_02*. Note that, although the bottom of both V_S and R connect to ground, this is electrically the same as for the circuit for Part 1.

3) Select the wattmeter tool and place the wattmeter to the right of resistor R.

4) Connect the wattmeter's voltage terminals across R.

5) Connect the wattmeter's current terminals between R and ground.

6) Double-click the wattmeter to expand it. Your circuit should look like that of Figure 3-5.

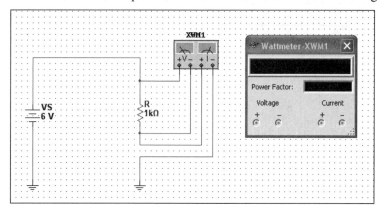

Figure 3-5: Circuit with Wattmeter Connected

7) Change the value of R to 510 Ω.

8) Change the value of V_S to 1.0 V.

9) Run the simulation.

10) Record the measured power to three significant digits as I_{MEAS} for V_S and R in Table 3-2.

11) Repeat Steps 6 and 7 for the remaining values of V_S.

12) Repeat Steps 5 through 8 for R = 1.0 kΩ.

13) Repeat Steps 5 through 8 for R = 2.2 kΩ.

Table 3-2: Calculated Power Values for Fixed Resistances

V_S (V)	$R = 510\ \Omega$		$R = 1.0\ \text{k}\Omega$		$R = 2.2\ \text{k}\Omega$	
	P_{CALC} (mW)	P_{MEAS} (mW)	P_{CALC} (mW)	P_{MEAS} (mW)	P_{CALC} (mW)	P_{MEAS} (mW)
1.0						
2.0						
3.0						
4.0						
5.0						
6.0						
7.0						
8.0						
9.0						
10.0						

3.8   The Power Curve

In this step, you will plot power versus voltage to create the power curve. This graph shows the power dissipated by a resistor for the any value of voltage across it. The shape of a *V-I* curve shows how the power dissipated by a resistor changes as the voltage across it changes.

1) For each value of V_S in Table 3-2, plot the corresponding value of P_{MEAS} for $R = 510\ \Omega$ on Plot 3-2. Label this graph "510 Ω".

2) For each value of V_S in Table 3-2, plot the corresponding value of P_{MEAS} for $R = 1\ \text{k}\Omega$ on Plot 3-2. Label this graph "1 kΩ".

3) For each value of V_S in Table 3-2, plot the corresponding value of P_{MEAS} for $R = 2.2\ \text{k}\Omega$ on Plot 3-2. Label this graph "2.2 kΩ".

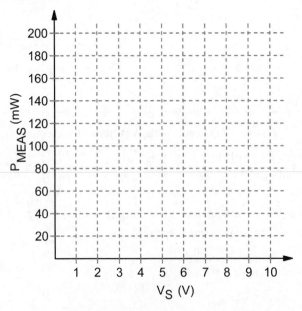

Plot 3-2: Measured Power Curves for Resistors

Observations:

Conclusions for Part 2

Questions for Part 2

1) From your power curve for the 510 Ω resistor, what is the approximate power dissipation when $V_S = 4.5$ V?

2) What is the voltage across the 2.2 kΩ resistor if the power dissipation is 20 mW?

3) A resistor dissipates 0.25 W when its voltage is 1.0 V. The voltage across it increases to 5.0 V. What is the new power dissipation?

Name _____   Class _____

Date _____   Instructor_____

DC/AC Experiment 4 - Series Resistive Circuits

4.1   Introduction

There are many ways to connect components in a circuit. One way is the **series** connection. A series connection is one in which only one current path exists between two points. A series circuit is a series connection of two or more components across an applied voltage. Because only one current path exists in a series circuit, the current through each component is the same. If the measured current for one component in a series circuit is 10 mA, the current through every other component in the circuit is also 10 mA. Another characteristic of a series circuit is that if any component opens (its resistance is infinite), no current can flow anywhere in the circuit. An example of a series circuit is an old-fashioned string of Christmas lights. If any bulb in the string burns out, current cannot flow and every other light in the string will also go out. The only way to fix the problem is to locate and replace the burnt-out bulb.

When resistors are connected in series, the total circuit resistance is the sum of the individual resistances. In Part 1 of this experiment, you will calculate the total resistance of each circuit and use Ohm's law to calculate the total series current. You will then use the DMM to measure the circuit current and resistance to verify your calculations.

An electronics law that is introduced with series circuits is **Kirchhoff's voltage law**. This law states that the algebraic sum (signed addition) of voltages in any closed loop is zero. Another version of this law is that the total voltage rise and total voltage drop for any closed loop must be equal. An analogy to this is the change in elevation while hiking through the mountains. Some parts of the hike are uphill and other parts are downhill, but if you end up back where you started the net change is elevation is zero. In Part 2 of this experiment, you will investigate and verify Kirchhoff's voltage law for series resistive circuits.

4.2   Reading

Floyd and Buchla, *Electric Circuits Fundamentals, 8th Ed.*, Chapter 4.

4.3   Key Objectives

Part 1:   Verify that the total resistance of a series circuit is the sum of the individual resistances.

Part 2:   Verify Kirchhoff's voltage law for series resistive circuits.

4.4   Multisim Circuits

Part 1:   *DC_AC_Exp_04_Part_01a, DC_AC_Exp_04_Part_01b*, and *DC_AC_Exp_04_Part_01c*.

Part 2:   *DC_AC_Exp_04_Part_02*

Part 1:   Total Series Resistance

1) Open the file *DC_AC_Exp_04_Part_01a*.

2) Visually verify that the circuit is a series circuit by confirming that only one path exists for current to flow from one terminal of V_S to the other.

3) Enter the values of V_S and resistors R_1 through R_5 for the circuit in Table 4-1.

4) Add the values of resistors R_1 through R_5 and enter this sum as $R_{T\,(CALC)}$ in Table 4-1.

5) Use Ohm's law to calculate the total current from V_S and R_T. Enter this value as $I_{T\,(CALC)}$ in Table 4-1.

6) Connect a DMM between V_S and R_1 and set it for dc ammeter mode.

7) Run the simulation.

8) Enter the measured current as $I_{T\,(MEAS)}$ in Table 4-1.

9) Stop the simulation.

10) Disconnect V_S from the circuit.

11) Connect the DMM across resistors R_1 through R_5 and set it for ohmmeter mode.

12) Run the simulation.

13) Enter the measured resistance as $R_{T\,(MEAS)}$ in Table 4-1.

14) Repeat steps 2 through 13 for the file *DC_AC_Exp_04_Part_01b*.

15) Repeat steps 2 through 13 for the file *DC_AC_Exp_04_Part_01c*.

Table 4-1: Series Resistor Circuit Measurements

Circuit	V_S	R_1	R_2	R_3	R_4	R_5	$R_{T\,(CALC)}$	$R_{T\,(MEAS)}$	$I_{T\,(CALC)}$	$I_{T\,(MEAS)}$
Part_01a										
Part_01b										
Part_01c										

Observations:

Conclusions for Part 1

Questions for Part 1

1) Would rearranging the order of resistors in the circuit change the total resistance or current? Why or why not?

2) Assume that when you prototype the circuit in *DC_AC_Exp_04_Part_01b* and measure the total resistance of the circuit from steps 10 to 13, you find that the DMM measures 25 kΩ. What is the most likely problem?

Part 2: Kirchhoff's Voltage Law

In this part of the experiment you must connect the DMM so that it displays the correct polarity of the measured voltage. To ensure that you connect the multimeter properly, always connect the multimeter to the circuit so that its positive (+) lead is consistently clockwise or counterclockwise from the negative (−) lead.

1) Open the file *DC_AC_Exp_04_Part_02*.

2) Connect a DMM across V_S and set the meter for dc voltmeter mode.

3) Run the simulation.

4) Record the measured voltage as V_S in the "Original" row of Table 4-2.

5) Measure and record the voltages for R_1 through R_5 in the V_{R1} through V_{R5} columns for the "Original" row of Table 4-2. You should observe that the signs of the measured voltages are opposite that for V_S.

6) Stop the simulation.

7) Add the values of V_S and V_{R1} through V_{R5} and record the sum in the "Net Voltage" column for the "Original" row of Table 4-2.

8) Double-click on R_1 to open its properties window.

9) Change the value in the **Resistance (R):** text box from "82 Ω" to "182 Ω".

10) Click the **OK** button to accept the change.

11) Increase the values of R_2 through R_5 by 100 Ω.

12) Run the simulation.

13) Record the measured voltage in the V_S column for the "Modified" row of Table 4-2.

14) Measure and record the voltages for R_1 through R_5 in the V_{R1} through V_{R5} columns for the "Modified" row of Table 4-2.

15) Stop the simulation.

16) Add the values of V_S and V_{R1} through V_{R5} and record the sum in the "Net Voltage" column for the "Modified" row of Table 4-2.

Table 4-2: Source and Resistor Voltage Measurements

Circuit	V_S (V)	V_{R1} (V)	V_{R2} (V)	V_{R3} (V)	V_{R4} (V)	V_{R5} (V)	Net Voltage (V)
Original							
Modified							

Observations:

Conclusions for Part 2

Questions for Part 2

1) Did changing the value of resistances in the circuit affect the value of net voltage? Why or why not?

2) Why must you consider the polarity of measured voltages when applying Kirchhoff's voltage law?

DC/AC Experiment 5 - Parallel Resistive Circuits

5.1   Introduction

A series connection is one in which only one current path exists between two points, so that the same current flows through each component. A parallel connection is one in which more than one current path exists between two points. A parallel circuit is a parallel connection of two or more components across an applied voltage. Because each component connects across the same two points, the voltage across each component is the same. If the measured voltage for one component in a parallel circuit is 5 V, the voltage across every other component in the circuit is also 5 V. A characteristic of a parallel circuit is that if any component opens (its resistance is infinite), current can still exist in the other parallel components. An example of a parallel circuit is the headlights in an automobile. If either headlamp burns out or is removed, current is still present in the other headlamp so that it is still lit.

In Part 1 of this experiment, you will investigate the characteristics of parallel resistive circuits. You will use Ohm's law to calculate the current through the parallel resistors as you add them to the circuit and use the DMM to measure the change in circuit current.

An electronics law that relates to parallel circuits is **Kirchhoff's current law**. This law states that the algebraic sum (signed addition) of currents at any node (point where two or more components connect) is zero. Another version of this law is that the total current entering a node and total current exiting the node must be equal. An analogy to this is water flowing through the pipes of an irrigation system. If ten gallons of water are entering the system every minute, then ten gallons must be leaving, as the pipes cannot expand to store the water and the water must go somewhere. In Part 2 of this experiment, you will verify Kirchhoff's current law for parallel resistive circuits and use the law to determine how to calculate the total resistance of parallel circuits.

5.2   Reading

Floyd and Buchla, *Electric Circuits Fundamentals, 8th Ed.*, Chapter 5.

5.3   Key Objectives

Part 1:   Work with groups of Multisim objects to explore the characteristics of parallel resistive circuits.

Part 2:   Verify Kirchhoff's current law for parallel resistive circuits and derive and verify the formula for the total resistance of parallel circuits.

5.4   Multisim Files

Part 1:   *DC_AC_Exp_05_Part_01*

Part 2:   *DC_AC_Exp_05_Part_02*

Part 1:   Characteristics of Parallel Resistive Circuits

1)   Open the Multisim file *DC_AC_Exp_05_Part_01*.

2)   Use Ohm's law to calculate the value of the current I_{R1} through R_1. Record this value in the $I_{R\,(CALC)}$ column for the R_1 row of Table 5-1.

3)   Run the simulation.

4)   Record the measured total current in the $I_{T\,(MEAS)}$ column for the R_1 row of Table 5-1.

5)   Use Ohm's law to calculate the total resistance from V_S and $I_{T\,(MEAS)}$. Record this value in the $R_{T\,(CALC)}$ column for the R_1 row of Table 5-1.

6)   Stop the simulation.

7)   Select R_1 and the ground symbol below it. There are two ways to do so.

   a.   Left-click on the workspace above and to the left of R_1. While holding the left mouse button down, drag the cursor down and to the right. As you do so, a rectangular selection window will appear.

Release the left mouse button when the selection window includes both R_1 and the ground symbol below it. The rectangle will disappear and dashed boxes will appear around R_1 and the ground symbol, showing that they are selected.

b. SHIFT + CLICK (hold down the "SHIFT" key and click the left mouse button) first on R_1 and then on the ground symbol below it. As you do so a dashed box will appear around each selected item. If you accidentally select an object that you do not want, SHIFT + CLICK the object again to de-select it. SHIFT + CLICK essentially toggles the selection status of Multisim objects: it will select objects that are not yet selected and de-select objects that are selected.

8) Copy the selected objects. You can do so in one of three ways.

    a. Select <u>Copy</u> from the **Edit** menu.

    b. Enter "CTRL + C" (hold down the "CTRL" key and press the "C" key).

    c. Select the **Copy** tool in the **Standard Toolbar** (see Figure 5-1).

Figure 5-1: Multisim Copy Tool

9) Paste the copied objects into the workspace. You can do so in one of three ways.

    a. Select <u>Paste</u> from the **Edit** menu.

    b. Enter "CTRL + V" (hold down the "CTRL" key and press the "V" key).

    c. Select the **Paste** tool in the **Standard Toolbar** (see Figure 5-2).

Figure 5-2: Multisim Paste Tool

Position the objects to the right of R_1, and left-click to paste them into the workspace.

10) Connect the top of the new resistor (note that Multisim automatically renumbers it as R_2) to the wire at the top of R_1. Your circuit should now be similar to Figure 5-3.

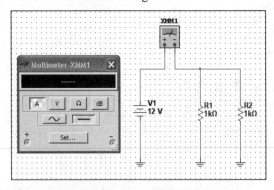

Figure 5-3: Circuit with R_2 Added

11) Use Ohm's law to calculate the value of the current I_{R2} through R_2. Record this value in the $I_{R\,(CALC)}$ column for the R_2 row of Table 5-1.

12) Run the simulation.

13) Record the measured total current in the $I_{T\,(MEAS)}$ column for the R_2 row of Table 5-1.

14) Use Ohm's law to calculate the total resistance from V_S and $I_{T\,(MEAS)}$. Record this value in the $R_{T\,(CALC)}$ column for the R_2 row of Table 5-1.

15) Calculate the change in total current from the previous value. Record this value in the ΔI_T column for the R_2 row of Table 5-1.

16) Stop the simulation.

17) Repeat steps 7 through 16 to add resistors R_3 through R_5 to the circuit and complete the R_3 through R_5 rows of Table 5-1.

Table 5-1: Data for Parallel 1 kΩ Resistor Circuit

Resistor Added	$I_{R\,(CALC)}$ (mA)	$I_{T\,(MEAS)}$ (mA)	$R_{T\,(CALC)}$	ΔI_T (mA)
R_1				
R_2				
R_3				
R_4				
R_5				

Observations:

Conclusions for Part 1

Questions for Part 1

1) What happens to the total current as more parallel resistors are added to the circuit?

2) What happens to the total resistance as more parallel resistors are added to the circuit?

Part 2:   Kirchhoff's Current Law

Kirchhoff's current law states that the total current entering a node must equal the total current leaving that node. For a parallel resistive circuit this means that the total current I_T from the voltage source must equal the total current through the parallel resistors, or

$$I_T = I_{R1} + I_{R2} + \dots + I_{RN}$$

Each resistor has the source voltage across it, so from Ohm's law for current

$$V_S / R_T = (V_S / R_1) + (V_S / R_2) + \dots + (V_S / R_N)$$
$$= V_S [(1 / R_1) + (1 / R_2) + \dots + (1 / R_N)]$$

Divide both sides by V_S.

$$1 / R_T = (1 / R_1) + (1 / R_2) + \dots + (1 / R_N)$$

Finally, take the reciprocal of both sides.

$R_T = 1 / [(1 / R_1) + (1 / R_2) + ... (1 / R_N)]$

Therefore, the total resistance of a parallel resistive circuit is the reciprocal of the sum of the reciprocals of the individual resistors. In this part of the experiment, you will verify that the equation for total parallel resistance and Kirchhoff's current law are valid.

1) For each of the indicated circuits in Table 5-2,

    a. Calculate the currents I_{R1} through I_{R4} for the indicated values of V_S and R_1 through R_4.

    b. Add the values of I_{R1} through I_{R4} and record this value as $I_{T(CALC)}$.

    c. Use Ohm's law to calculate the value of R_T and record this value as $R_{T(CALC)}$.

    d. Calculate R_T from the indicated values of R_1 through R_4 and record this value as $R_{T(CALC)}$.

2) Open the Multisim file *DC_AC_Exp_05_Part_02*.

3) For each of the indicated circuits in Table 5-2,

    a. Change the values of V_S and R_1 through R_4 to the indicated values.

    b. Simulate the circuit.

    c. Record the measured current as $I_{T(MEAS)}$.

    d. Stop the simulation.

    e. Disconnect V_S from the circuit.

    f. Set the DMM to ohmmeter mode and connect it between the top of the parallel resistors and ground.

    g. Start the simulation.

    h. Record the measured resistance value as $R_{T(MEAS)}$.

    i. Stop the simulation.

Table 5-2: Parallel Resistive Circuit Parameters

Parameter	Circuit 1	Circuit 2	Circuit 3	Circuit 4
V_S	6.0 V	10.0 V	12.0 V	24.0 V
R_1	1.0 kΩ	2.2 kΩ	15 kΩ	100 kΩ
R_2	1.0 kΩ	4.7 kΩ	33 kΩ	390 kΩ
R_3	1.0 kΩ	5.1 kΩ	68 kΩ	750 kΩ
R_4	1.0 kΩ	10 kΩ	91 kΩ	1.0 MΩ
I_{R1}				
I_{R2}				
I_{R3}				
I_{R4}				
$I_{T(CALC)}$				
R_T				
$R_{T(CALC)}$				
$I_{T(MEAS)}$				
$R_{T(MEAS)}$				

Observations:

Conclusions for Part 2

Questions for Part 2

1) From the data in Table 5-2, how does the total parallel resistance compare with the smallest parallel resistor value?

2) You connect a 100 Ω resistor in parallel with an unknown resistance. What can you say for certain about the total parallel resistance?

3) What is the total parallel resistance for a circuit if $R_1 = 2.4$ kΩ, $R_2 = 1.8$ kΩ, $R_3 = 5.6$ kΩ, and $R_4 = 4.3$ kΩ?

DC/AC Experiment 6 - Series-Parallel Resistive Circuits

6.1 Introduction

Very few electronic circuits are strictly series circuits or parallel circuits. The vast majority are some combination of the two types. Series-parallel circuit analysis consists of identifying the series and parallel connections and then working through the circuit by applying the appropriate analysis technique to part of the circuit. Special circuit notation is often used to represent how components are connected and help simplify circuit calculations. A common convention uses "+" to represent a series connection, and "||" to represent a parallel connection. As an example, "$R_1 + R_2$" means that R_1 and R_2 are in series, while "$R_1 \| R_2$" means that R_1 and R_2 are in parallel. The standard order of operations evaluates the "||" operator before the "+" operator, although parentheses can be used to alter this to reflect how the circuit is connected. Refer to Figure 6-1 to see the difference between the circuits that correspond to "$R_1 + R_2 \| R_3 + R_4$", "$(R_1 + R_2) \| R_3 + R_4$", and $(R_1 + R_2) \| (R_3 + R_4)$".

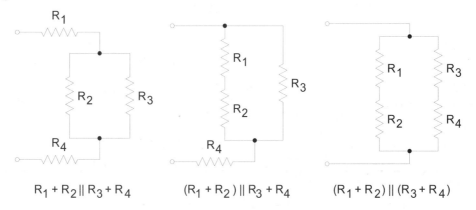

$$R_1 + R_2 \| R_3 + R_4 \qquad (R_1 + R_2) \| R_3 + R_4 \qquad (R_1 + R_2) \| (R_3 + R_4)$$

Figure 6-1: Circuit Notation Examples

When analyzing series-parallel circuits, the simplest approach is often to work through the expression for the circuit to find the total resistance, progressively replacing the series and parallel combinations with the calculated equivalent resistance for each combination, and then work backwards using Ohm's law to determine the individual currents and voltages in the circuit. For the circuit represented by $(R_1 + R_2) \| R_3 + R_4$, for example, the value of the total resistance R_T is $R_T = (R_1 + R_2) \| R_3 + R_4$. The value for the first equivalent resistance R_{EQ1} would be $R_{EQ1} = R_1 + R_2$, as the series combination $(R_1 + R_2)$ is evaluated first. Once R_{EQ1} is known, the circuit simplifies to $R_T = R_{EQ1} \| R_3 + R_4$. The value of the second equivalent resistance R_{EQ2} would then be $R_{EQ2} = R_{EQ1} \| R_3$, as the parallel combination $R_{EQ1} \| R_3$ is evaluated next. Once R_{EQ2} is known, the circuit simplifies to $R_T = R_{EQ2} + R_4$. The final equivalent resistance R_{EQ3} would then be $R_{EQ3} = R_{EQ2} + R_4$, as the series combination of R_{EQ2} and R_4 is evaluated last. Once R_{EQ3} is known, the circuit simplifies to $R_T = R_{EQ3}$. Once you know R_T, you can then use Ohm's law to determine the total current I_T and work backwards through the circuit to find the voltages and currents for each resistor.

In Part 1 of this experiment, you will calculate the resistance, voltages, and currents for typical series-parallel circuits. In Part 2, you will use Multisim to verify your calculations.

6.2 Reading

Floyd and Buchla, *Electric Circuits Fundamentals, 8th Ed.*, Chapter 6.

6.3 Key Objectives

Part 1: Use series and parallel notation to represent series-parallel resistive circuits and calculate the total resistance, resistor voltages, and resistor currents.

Part 2: Use Multisim to verify calculations from Part 1.

6.4   Multisim Files

Part 2:   *DC_AC_Exp_06_Part_02a*, *DC_AC_Exp_06_Part_02b*, and *DC_AC_Exp_06_Part_02c*.

Part 1:   Series-Parallel Circuit Representation and Calculations

6.5   Series Resistors in Parallel

For this part of the experiment, refer to the circuit in Figure 6-2.

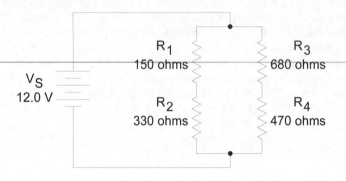

Figure 6-2: Series Resistors in Parallel

1) What is the expression for total resistance for the circuit?

$R_T = (R_1 + R_2) \parallel (R_3 + R_4)$

2) Record the values of V_S, V_{REQ1}, V_{REQ2}, and R_1 through R_4 in Table 6-1.

3) Use R_1 through R_4 to determine R_{EQ1}, R_{EQ2}, and R_T. Record these values in Table 6-1.

4) Use V_S and R_T to calculate I_T. Record this value in Table 6-1.

5) Use V_{REQ1} and V_{REQ2} to calculate I_{REQ1} and I_{REQ2}. Record these values in Table 6-1.

6) Use I_{REQ1}, I_{REQ2}, and R_1 through R_4 to calculate V_{R1} through V_{R4} and I_{R1} through I_{R4}. Record these values in Table 6-1.

Table 6-1: Calculated Values for Series Resistors in Parallel

Parameter	Value	Parameter	Value	Parameter	Value
V_S		$V_{REQ1} = V_S$		$V_{REQ2} = V_S$	
R_1		$I_{REQ1} = V_{REQ1} / R_{EQ1}$		$I_{REQ2} = V_{REQ2} / R_{EQ2}$	
R_2		$I_{R1} = I_{REQ1}$		$I_{R3} = I_{REQ1}$	
R_3		$I_{R2} = I_{REQ1}$		$I_{R4} = I_{REQ1}$	
R_4		$V_{R1} = I_{R1} R_1$		$V_{R3} = I_{R3} R_3$	
$R_{EQ1} = R_1 + R_2$		$V_{R2} = I_{R2} R_2$		$V_{R4} = I_{R4} R_4$	
$R_{EQ2} = R_3 + R_4$					
$R_T = R_{EQ1} \parallel R_{EQ2}$					
$I_T = V_S / R_T$					

6.6   Parallel Resistors in Series

For this part of the experiment, refer to the circuit in Figure 6-3.

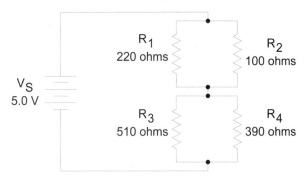

Figure 6-3: Parallel Resistors in Series

1)   What is the expression for total resistance for the circuit?

$R_T =$

2)   Record the values of V_S and R_1 through R_4 in Table 6-2.

3)   Use R_1 through R_4 to determine R_{EQ1}, R_{EQ2}, and R_T. Record these values in Table 6-2.

4)   Use V_S and R_T to calculate I_T. Record this value in Table 6-2.

5)   Use I_T, R_{EQ1} and R_{EQ2} to calculate I_{REQ1}, I_{REQ2}, V_{REQ2}, and V_{REQ2}. Record these values in Table 6-2.

6)   Use V_{REQ1}, V_{REQ2}, and R_1 through R_4 to calculate V_{R1} through V_{R4} and I_{R1} through I_{R4}. Record these values in Table 6-2.

Table 6-2: Calculated Values for Parallel Resistors in Series

Parameter	Value	Parameter	Value	Parameter	Value
V_S		$I_{REQ1} = I_T$		$I_{REQ2} = I_T$	
R_1		$V_{REQ1} = I_{REQ1} R_{EQ1}$		$V_{REQ2} = I_{REQ2} R_{EQ2}$	
R_2		$V_{R1} = V_{REQ1}$		$V_{R3} = V_{REQ2}$	
R_3		$V_{R2} = V_{REQ1}$		$V_{R4} = V_{REQ2}$	
R_4		$I_{R1} = V_{R1} / R_1$		$I_{R3} = V_{R3} / R_3$	
$R_{EQ1} = R_1 \| R_2$		$I_{R2} = V_{R2} / R_2$		$I_{R4} = V_{R4} / R_4$	
$R_{EQ2} = R_3 \| R_4$					
$R_T = R_{EQ1} + R_{EQ2}$					
$I_T = V_S / R_T$					

6.7   The Ladder Circuit

For this part of the experiment, refer to the circuit in Figure 6-4.

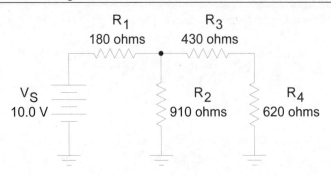

Figure 6-4: Ladder Circuit

1) What is the expression for total resistance for the circuit?

$R_T =$

2) Record the values of V_S and R_1 through R_4 in Table 6-3.
3) Use R_1 through R_4 to determine R_{EQ1}, R_{EQ2}, and R_T in Table 6-3.
4) Use V_S and R_T to calculate I_T in Table 6-3.
5) Use I_T, R_{EQ1}, R_{EQ2}, and R_1 through R_4 to calculate I_{REQ1}, I_{REQ2}, V_{REQ1}, V_{REQ2}, V_{R1} through V_{R4}, and I_{R1} through I_{R4} in Table 6-3.

Table 6-3: Calculated Values for Ladder Circuit

Parameter	Value	Parameter	Value	Parameter	Value
V_S		$I_{R1} = I_T$		$V_{REQ1} = V_{REQ2}$	
R_1		$V_{R1} = I_{R1} R_1$		$I_{REQ1} = V_{REQ1} / R_{EQ1}$	
R_2		$I_{REQ2} = I_T$		$I_{R3} = I_{REQ1}$	
R_3		$V_{REQ2} = I_{REQ2} R_{EQ2}$		$I_{R4} = I_{REQ1}$	
R_4		$V_{R2} = V_{REQ2}$		$V_{R3} = I_{R3} R_3$	
$R_{EQ1} = R_3 + R_4$		$I_{R2} = V_{R2} / R_2$		$V_{R4} = I_{R4} R_4$	
$R_{EQ2} = R_2 \parallel R_{EQ1}$					
$R_T = R_1 + R_{EQ2}$					
$I_T = V_S / R_T$					

Conclusions for Part 1

Questions for Part 1

1) For the data in Table 6-1, what circuit property justifies setting V_{REQ1} and V_{REQ2} equal to V_S?

2) For the data in Table 6-2, what circuit property justifies setting I_{REQ1} and I_{REQ2} equal to I_T?

3) For the data in Table 6-3, do the currents through R_1, R_2, and R_3 satisfy Kirchhoff's current law?

4) For the data in Table 6-3, do the voltages across R_2, R_3, and R_4 satisfy Kirchhoff's voltage law?

Part 2:   Series-Parallel Circuit Measurements

6.8   Series Resistors in Parallel

1) Open the Multisim file *DC_AC_Exp_06_Part_02a*.

2) Measure and record to three significant digits each of the parameters in Table 6-4.

Table 6-4: Measurements for Series Resistors in Parallel Circuit

Parameter	Value	Parameter	Value	Parameter	Value
I_T		V_{R1}		I_{R1}	
R_T		V_{R2}		I_{R2}	
		V_{R3}		I_{R3}	
		V_{R4}		I_{R4}	

6.9   Parallel Resistors in Series

1) Open the Multisim file *DC_AC_Exp_06_Part_02b*.

2) Measure and record to three significant digits each of the parameters in Table 6-5.

Table 6-5: Measurements for Parallel Resistors in Series Circuit

Parameter	Value	Parameter	Value	Parameter	Value
I_T		V_{R1}		I_{R1}	
R_T		V_{R2}		I_{R2}	
		V_{R3}		I_{R3}	
		V_{R4}		I_{R4}	

6.10   The Ladder Circuit

1) Open the Multisim file *DC_AC_Exp_06_Part_02c*.

2) Measure and record to three significant digits each of the parameters in Table 6-6.

Table 6-6: Measurements for Ladder Circuit

Parameter	Value	Parameter	Value	Parameter	Value
I_T		V_{R1}		I_{R1}	
R_T		V_{R2}		I_{R2}	
		V_{R3}		I_{R3}	
		V_{R4}		I_{R4}	

Conclusions for Part 2

Questions for Part 2

1) How do the measured values for Table 6-4 compare with the calculated values for Table 6-1?

2) How do the measured values for Table 6-5 compare with the calculated values for Table 6-2?

3) How do the measured values for Table 6-6 compare with the calculated values for Table 6-3?

4) Open the Multisim circuit *DC_AC_Exp_06_Part_02a*, replace R_1 and R_2 with your calculated value of R_{EQ1}, and measure the total current. Does the total current differ from that for the original circuit?

5) Open the Multisim circuit *DC_AC_Exp_06_Part_02b*, replace R_3 and R_4 with your calculated value of R_{EQ2}, and measure the total current. Does the total current differ from that for the original circuit?

6) Open the Multisim circuit *DC_AC_Exp_06_Part_02c*, replace R_2, R_3, and R_4 with your calculated value of R_{EQ2}, and measure the total current. Does the total current differ from that for the original circuit?

7) What do your findings for questions 4 through 6 indicate?

Name _____   Class _____

Date _____   Instructor_____

DC/AC Experiment 7 - Relays

7.1   Introduction

Electricity and magnetism are very closely linked, so much so that they are commonly studied under the single topic of electromagnetism. Moving charge creates a magnetic field, and magnetic fields affect the movement of charge. In some cases this relationship between electricity and magnetism can create problems, but it is also essential for some electric components and circuits to operate. Components that rely on and use magnetism as part of their operation include solenoids, relays, magnetic reed switches, and transformers.

In this experiment you will study the operation of relays and some relay applications. A basic relay consists of a coil and a moving armature. When current flows through the coil the electromagnetic field pulls the armature towards the coil, opening or closing one or more sets of contacts connected to the armature. A relay whose contacts are open when no current is through its coil is called a normally open relay, and a relay whose contacts are closed when no current is through the coil is called a normally closed relay. Refer to Figure 7-1.

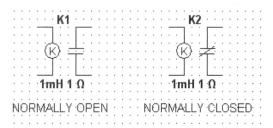

Figure 7-1: Normally Open and Normally Closed Relays

The current rating of the contacts can be much larger than the current through the coil. This allows a small current to control a much larger current, such as in the case of the starter relay in an automobile engine. While this is not actually an example of electronic amplification as with transistors or vacuum tubes, relays can reduce the power dissipated by the control circuitry in a system, thereby simplifying the system.

Relays were a fundamental unit in constructing the logic circuits in early electromechanical computers. A moth became caught in the contacts of a relay in the Harvard Mark II computer in 1947 and kept the computer from functioning properly. When technicians removed the insect from the machine, computer pioneer Grace Hopper remarked that they were "debugging" the computer and brought the terms "bug" and "debugging" into popular use.

In Part 1 of this experiment, you will examine the basic relay operation, measure the hysteresis for the coil current of a relay, and observe the effect of hysteresis on noise in a system. In Part 2, you will study a latching relay application, observe a phenomenon known as inductive kickback, and investigate an example of relay logic.

7.2   Reading

Floyd and Buchla, *Electric Circuits Fundamentals, 8th Ed.*, Chapter 7

7.3   Key Objectives

Part 1:   Measure the hysteresis for the coil current of a relay and observe its effect on noise in a circuit.

Part 2:   Verify the latching action of a relay circuit, observe the effects of inductive kickback, and investigate use of relays in implementing digital logic.

7.4   Multisim Files

Part 1:   *DC_AC_Exp_07_Part_01a* and *DC_AC_Exp_07_Part_01b*

Part 2:   *DC_AC_Exp_07_Part_02a* and *DC_AC_Exp_07_Part_02b*.

Part 1:   Relay Hysteresis

7.5   Measuring Hysteresis

1) Open the Multisim circuit *DC_AC_Exp_07_Part_01a*.

2) Start the simulation.

3) Hover the cursor over the potentiometer so that the slider control appears. Drag the slider to the left to reduce the potentiometer resistance in 1% increments, gradually increasing the coil current through the relay. Record the potentiometer setting and coil current for which the lamp first turns on.

4) Move the slider one increment to the right to decrease the coil current. Does the lamp turn off?

5) Drag the slider to the right to increase the potentiometer resistance in 1% increments, gradually reducing the coil current through the relay. Record the potentiometer setting and coil current for which the lamp first turns off.

6) Move the slide one increment to the left to increase the coil current. Does the lamp turn on?

7) Stop the simulation.

7.6   Effects of Hysteresis

1) Open the Multisim circuit *DC_AC_Exp_07_Part_01b*. This circuit is the same as the Multisim circuit *DC_AC_Exp_07_Part_01a*, except that the relay hysteresis has been reduced.

2) Start the simulation.

3) Hover the cursor over the potentiometer so that the slider control appears. Drag the slider to the left to reduce the potentiometer resistance in 1% increments, gradually increasing the coil current through the relay. Record the potentiometer setting for which the lamp first turns on and describe the behavior of the relay contacts and lamp.

4) Drag the slider further to the left in 1% increments until the relay contacts are continually closed and the lamp is continually on. Record the potentiometer setting.

5) Move the slide one increment to the right to decrease the coil current. What happens?

6) Stop the simulation.

Conclusions for Part 1

Questions for Part 1

1) What is the hysteresis (difference between the coil currents that close and open the relay contacts) for the relay in *DC_AC_Exp_07_Part_01a*?

2) What difference did the larger hysteresis for the coil current in *DC_AC_Exp_07_Part_01a* have compared to *DC_AC_Exp_07_Part_01b*?

Part 2: Relay Application

7.7 Latching Alarm

1) Open the Multisim circuit *DC_AC_Exp_07_Part_02a*. This circuit represents an alarm system that protects three access points to a house: the front door, the back door, and the patio door.

2) Start the simulation.

3) If the "ARM" switch is closed, press the space bar or double-click on the switch to open it.

4) Close each of the alarm sensor switches corresponding to the three protected doors. Does the "ALARM" lamp light up?

5) Open each of the alarm switches.

6) Close the "ARM" switch, simulating that you have armed the security system.

7) Close any of the alarm sensor switches, simulating an intruder entering the house, and describe what happens.

8) Open the alarm switch that you closed in step 7, simulating the intruder attempting to deactivate the alarm sensor at the point of entry, and describe what happens.

9) Open and close any of the other alarm sensor switches and describe what happens.

10) Open the "ARM" switch, simulating that you have deactivated the security system, and describe what happens.

11) Stop the simulation.

7.8 Inductive Kickback

When the resistance to current that is flowing in a conductive coil increases suddenly, such as when a switch opens, the inductance in the coil attempts to keep the same amount of current flowing by developing a voltage large enough to overcome the increased resistance. This phenomenon is known as **inductive kickback** and can result in very high voltages that could damage components in series with the coil or cause an electric arc. The flash you sometimes see when turning off a light switch or the spark plugs firing in an engine are examples of the latter. Relay circuits typically include a diode (which you will study in later experiments) or some other form of surge protection across the coil that can withstand the induced voltage and carry the current until the energy in the coil dissipates. The diode D_1 across the relay coil in *DC_AC_Exp_07_Part_02a* protects the alarm circuit from the effect of inductive kickback.

1) Open the Multisim circuit *DC_AC_Exp_07_Part_02a*.

2) Delete the diode D_1 and its ground from the circuit.

3) Close the "ARM" switch to arm the alarm system.

4) Close any of the alarm switches and describe what happens.

5) Close the alarm switch and describe what happens.

6) Open and close any of the alarm sensor switches and describe what happens.

7) Stop the simulation.

7.9 Relay Logic

Relay logic is rarely used today, although the ladder logic in modern programmable logic controllers (PLCs) is based on relay logic. In this part of the experiment you will investigate a circuit that implements the logic needed to activate the danger indicator for a lubricant pumping system. The system has two switches. When the pump switch is closed, the pump is running. When the pump switch is open, the pump is not running. When the valve switch is closed, the pressure relief valve is closed so that lubricant is kept in the system. When the valve switch is open, the pressure relief valve is open so that the lubricant drains from the system. Table 7-1 shows the combination of switch positions and the system condition for each combination.

Table 7-1: Pump System Switch Combinations

Pump Switch	Valve Switch	Condition	Status
CLOSED	CLOSED	Lubricant circulating	Safe
CLOSED	OPEN	Lubricant released from system	Danger
OPEN	CLOSED	Lubricant not circulating	Danger
OPEN	OPEN	Pumping system inactive	Safe

From the table, the danger indicator should be ON if the pump switch is closed AND the valve switch is open OR if pump switch is open AND the valve switch is closed. Refer to Figure 7-2.

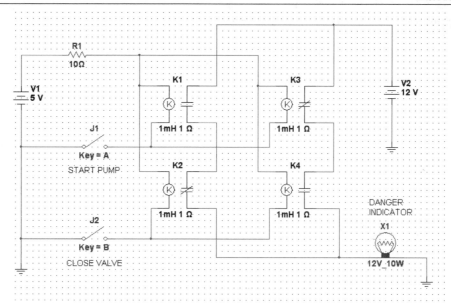

Figure 7-2: Pumping System Danger Indication Circuit

When the pump switch J_1 is closed, current through the coils of K_1 and K_3 closes the contacts of K_1 and opens the contacts of K_3. When the valve switch J_2 is closed, current through the coils of K_2 and K_4 opens the contacts of K_2 and closes the contacts of K_4. The contacts of K_1 and K_2 are in series and the contacts of K_3 and K_4 are in series, and both sets of relays are in parallel. Consequently, current from V_2 will light the lamp when the contacts of K_1 AND K_2 are both closed OR when the contacts of K_3 AND K_4 are both closed.

1) For each set of switch positions in Table 7-2, indicate whether the danger indicator lamp will be "ON" or "OFF" in the "Predicted" column.

2) Open the Multisim file *DC_AC_Exp_07_Part_02b*.

3) Start the simulation.

4) Use the "A" and "B" keys to set the switches J_1 and J_2 to each of the positions in Table 7-2. Record whether the lamp is "ON" or "OFF" in the "Observed" column.

5) Stop the simulation.

Table 7-2: Danger Indicator States

J_1	J_2	Predicted	Observed
OPEN	OPEN		
OPEN	CLOSED		
CLOSED	OPEN		
CLOSED	CLOSED		

Conclusions for Part 2

Questions for Part 2

1) Why are the alarm sensor switches in parallel rather than series?

2) How would the alarm system work if the alarm sensor switches were in series rather than parallel?

3) For the circuit in Figure 7-2, when would the danger indicator light up if K_1 was a normally closed relay and K_3 was a normally open relay?

Name _____   Class _____

Date _____   Instructor_____

DC/AC Experiment 8 - AC Measurements

8.1   Introduction

Up to now you have used the multimeter for your circuit measurements. The multimeter is ideal for making many types of precise circuit measurements, but its intent is to measure values that are static (i.e., values that do not significantly change over time). The instrument for measuring dynamic, or time-varying, values is the oscilloscope. An oscilloscope is generally less accurate than a digital multimeter and limited to voltage measurements but can display rapidly changing signals. Most oscilloscopes have multiple channels so that you can compare or combine signals, measure delays between events, and use an event on one channel to initiate the capture of information on another. Other common features of modern oscilloscopes include measurement cursors, digital time and amplitude displays, automatic detection of maximum and minimum voltages, provisions to upload waveforms to removable media or computers, and on-line help screens.

The Multisim software offers several varieties of oscilloscopes. The generic two- and four-channel oscilloscopes provide basic features that are common to most oscilloscopes. The other oscilloscopes emulate the form, features, and functions of actual oscilloscope models. Although the latter are excellent tools for learning to use the actual oscilloscopes, this experiment will use the two-channel generic oscilloscope to present the basic concepts and features that are common to all oscilloscopes. In later experiments you will use the simulated Tektronix scope that Multisim provides.

In Part 1, you will examine the typical features of an oscilloscope and demonstrate how to read an oscilloscope display. In Part 2, you will use the oscilloscope controls to observe, measure, and adjust the display of ac voltages.

8.2   Reading

Floyd and Buchla, *Electric Circuits Fundamentals, 8th Ed.*, Chapter 8

8.3   Key Objectives

Part 1:   Identify typical features of a two-channel oscilloscope, demonstrate the ability to read an oscilloscope display, and calculate ac voltage and frequency values.

Part 2:   Use the Multisim two-channel oscilloscope controls to observe, measure, and adjust the display of ac voltages.

8.4   Multisim Files

Part 2:   *DC_AC_Exp_08_Part_02a* and *DC_AC_Exp_08_Part_02b*

Part 1:   The Oscilloscope Controls

8.5   Accessing the Oscilloscope

To access the two-channel oscilloscope, click the **Oscilloscope** tool (refer to Figure 8-1) in the **Instruments Toolbar**.

Figure 8-1: Oscilloscope Tool

Figure 8-2 shows the minimized view of the two-channel oscilloscope. The oscilloscope has three circuit connections. The two connections on the bottom of the oscilloscope are for the A channel (left) and B channel (right) signal and ground. The connections on the right of the oscilloscope are for the external trigger signal and ground.

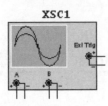

Figure 8-2: Minimized 2-Channel Oscilloscope View

Figure 8-3 shows the enlarged view of the two-channel oscilloscope with a sample display.

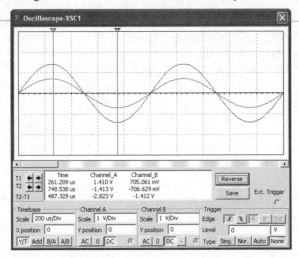

Figure 8-3: Enlarged Two-Channel Oscilloscope View

The oscilloscope controls consist of the following six sections:

- Graphical display
- Display controls
- Timebase controls
- Channel A controls
- Channel B controls
- Trigger controls

8.6   Graphical Display

The display occupies most of the upper portion of the enlarged oscilloscope view. This is the area in which you view the channel signals and position the reference cursors. Practical oscilloscope displays are 10 divisions wide by 8 divisions high, but the Multisim generic oscilloscope displays are 10 divisions wide by 6 divisions high. The display in Figure 8-3 shows two signals, one from Channel A and one from Channel B, and reference cursors 1 and 2 (Multisim refers to these as "crosshairs" 1 and 2 in the generic oscilloscope), which appear as vertical lines. The small triangle at the top of each cursor identifies the number of the cursor.

The reference cursors provide amplitude and time information for a specific point on the channel signals. The cursors are normally located at the far right and left sides of the display. To directly position a cursor, use the left mouse button to click and drag the cursor to the desired position in the display and release the left mouse button. You can also right-click the cursor and select a specific time or amplitude value to position the cursor.

Directly beneath the graphical display is a scroll bar. If the collected data extends beyond one screen, you can use the scroll bar to examine parts of the signal that are not on the screen.

8.7 Display Controls

The display controls are just below the oscilloscope display area. This section allows you to position the cursors, view the amplitude and time information for the channel signals at the cursor positions, and select the color of the display background. Refer to Figure 8-4.

Figure 8-4: Oscilloscope Display Controls

The ◆ and ➡ buttons to the right of the **T1** and **T2** labels adjust the position of reference cursors 1 and 2, respectively. The information associated with reference cursors 1 and 2 is in the window to the right of the **T1** and **T2** buttons, respectively. The amplitude and time information associated with each cursor will update as you adjust the position of the cursors.

The information in the window to the right of the **T2-T1** label is the difference in amplitude and time between reference cursor 2 and reference cursor 1. This feature allows you to easily measure time delays, signal periods, peak-to-peak amplitudes, and other differential data.

The **Reverse** button allows you to change the background of the graphical display to black or white to improve the visibility of the signals.

The **Save** button allows you to save the graphical display as a list of time and amplitude data points in a scope display (.SCP), LabVIEW measurement (.LVM), or TDM file format so that you can import the data into other applications. The .SCP and .LVM files are in text format that you can open with a number of text editors, word processors, and spreadsheets. In addition, Multisim's Grapher utility (under the **View** menu) can open the .SCP file and display it as a graphic that you can view, save, and print. The .TDM file format is in a binary file format that is compatible with National Instruments' DIAdem data management and analysis software.

8.8 Timebase Controls

The controls in the **Timebase** section allow you to adjust the horizontal position and scale of the display and select the format of the display. Refer to Figure 8-5.

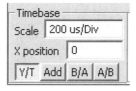

Figure 8-5: Oscilloscope Timebase Controls

The **Scale** value specifies how much time each horizontal division represents. The time settings use the standard 1-2-5 progression so that each setting is about twice that of the previous setting. For example, the setting before the 200 μs/Div setting is 100 μs/Div and the setting after it is 500 μs/Div. You can use the up and down scroll buttons to set this value from 1 fs (10^{-15} seconds) per division to 1000 Ts (10^{12} seconds) per division. The 1 fs setting is one million times shorter than any practical scope can achieve. Note also that, unless you have nothing else to do for a while, you should avoid using the 1000 Ts/div setting, as each horizontal division at this setting equals approximately 31.7 million years!

The **X position** value allows you to manually shift the display in 0.1-division increments or use the scroll arrows to shift the display in 0.2-division increments to the left or right. This allows you to better align the display with a specific point on the horizontal axis.

The four buttons at the bottom of the **Timebase** section allow you to choose the format of the graphical display.

The **Y/T** button configures the oscilloscope to display the Channel A and Channel B signals separately with the vertical axis configured for volts and the horizontal axis configured for time. This is the typical operating mode of oscilloscopes.

The **Add** button configures the oscilloscope to add the Channel A and Channel B signals and display the result as a single signal with the vertical axis configured for volts and the horizontal axis configured for time. The mode is useful for finding the voltage across a component that has no direct connection to circuit ground.

The **B/A** button configures the oscilloscope to plot the Channel A signal against the horizontal axis and the Channel B signal against the vertical axis to create a two-dimensional plot called a Lissajous figure. This is a convenient display mode for determining the relative amplitude, frequency, and phase of two signals. Both the vertical and horizontal axes are configured for amplitude although the oscilloscope shows no units.

The function of the **A/B** button is similar to the **B/A** button, except that the oscilloscope plots the Channel B signal against the horizontal axis and the Channel A signal against the vertical axis.

Note that the **Scale** and **X position** controls will work only with the **Y/T** and **Add** modes.

8.9   Channel A Controls

The controls in the **Channel A** section allow you to adjust the vertical position and scale of the Channel A signal. Refer to Figure 8-6.

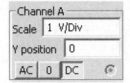

Figure 8-6: Oscilloscope Channel A Controls

The **Scale** value specifies how many volts each vertical division represents. The voltage settings use the standard 1-2-5 progression. For example, the setting before the 1 V/Div setting is 500 mV/Div, and the setting after it is 2 V/Div. You can use the up and down scroll buttons to set this value from 1 pV per division to 1000 TV per division. Just for reference, 1000 TV is enough electrical potential to generate a lightning bolt 189,000 miles long. If you plan to measure voltages on this order of magnitude, be sure to observe adequate safety precautions!

The **Y position** value allows you to manually shift the display in 0.1-division increments or use the scroll arrows to shift the display in 0.2-division increments up or down. This allows you to separate the Channel A and B signals for better viewing or to compensate for some unwanted dc offset in the signal.

The **AC**, **0**, and **DC** buttons determine the signal coupling for the channel.

The **AC** button removes any dc offset from the signal, so that Channel A couples (allows in) only the ac portion of the signal into the oscilloscope.

The **0** button (usually called GND on practical scopes) disconnects Channel A from the probe signal and directly to ground. This is useful if you want to determine a 0 V reference for a signal on Channel A or if you wish to view only the B channel when the oscilloscope is in the **Add** mode.

The **DC** button couples both the ac and dc components of the signal on Channel A into the oscilloscope. You will often require this coupling when you are viewing low-frequency signals so that the oscilloscope does not attenuate the signal.

8.10   Channel B Controls

The controls in the **Channel B** section are identical to those in the Channel A section, with the addition of one more button. Refer to Figure 8-7.

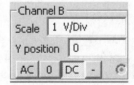

Figure 8-7: Oscilloscope Channel B Controls

The extra button, marked with a "–", inverts the signal on Channel B. You typically use this button when you wish to find the difference between the Channel A and B signals, such as when you wish to measure the voltage across an ungrounded, or "floating", component. To do this, select the **Add** mode and then activate the Channel B "–" button. Because this will invert the Channel B signal, the oscilloscope will display the waveform of "Channel A – Channel B", rather than "Channel A + Channel B".

8.11   Trigger Controls

The controls in the **Trigger** section determine the conditions that will trigger the oscilloscope (that is, cause the oscilloscope to display waveforms). Refer to Figure 8-8.

Figure 8-8: Oscilloscope Trigger Control

The **Edge** controls specify whether the trigger voltage must be increasing (a rising edge) or decreasing (a falling edge) for the oscilloscope to display the Channel A and Channel B signals. Refer to Figure 8-9, which specifies a rising-edge trigger, also called a leading-edge or positive-edge trigger. This means that the trigger voltage must exceed the **Level** value to trigger the oscilloscope. A falling-edge trigger, also called a trailing-edge or negative-edge trigger, means that the trigger voltage must fall below the **Level** value to trigger the oscilloscope.

Figure 8-9: Rising Edge Trigger

The **A**, **B**, and **Ext** buttons specify whether the oscilloscope uses the signal on Channel A, Channel B, or External Trigger for the trigger voltage.

The **Level** value sets the voltage level for the trigger signal. You can use the scroll buttons to specify the value of the trigger signal or manually enter the value in the text box. Click in the units box and select the unit you wish to use for the trigger level from the list.

The **Type** buttons determine the type of triggering.

The **Sing.** (single-sweep) button configures the oscilloscope to make a single sweep when the oscilloscope receives a valid trigger. After the oscilloscope completes a sweep across the screen, it should halt until you use one of the **Type** buttons to initiate a new sweep. In actuality the scope will continue to capture waveforms, although the display will continue to show the date from the first screen.

The **Nor.** (normal) button is similar to the single-sweep button, but after the oscilloscope completes a sweep across the screen it will clear the screen and initiate a new sweep if it receives a valid trigger.

The **Auto** (auto-trigger) button initiates a sweep whenever either of the following events occurs:

1)   The oscilloscope receives a valid trigger.

2)   A pre-defined amount of time has passed and the oscilloscope has not received a valid trigger.

The **None** button specifies that there are no specific trigger conditions.

Most applications will use the auto-trigger mode, although nonperiodic signals or special conditions can require other trigger modes for best results.

8.12   Oscilloscope Measurement Terminology

A static value possesses only a single characteristic, namely magnitude or amplitude that describes it. Time-varying signals have both time and amplitude characteristics to describe them. When you use an oscilloscope to observe a time-varying signal, you will measure specific amplitude and time characteristics for the signal.

The most common signal you will observe in ac electronics is the sine wave. Refer to Figure 8-10.

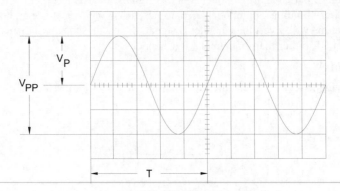

Figure 8-10: Example of Sine Wave Display

V_{pp} is the peak-to-peak voltage. The peak-to-peak voltage for a sine wave is the difference between the minimum and maximum amplitudes. V_{pp} for the sine wave in Figure 8-10 is four divisions.

V_p is the peak voltage, which is half the peak-to-peak value for a sine wave. V_p for the sine wave in Figure 8-10 is two divisions.

T is the period of the sine wave, which is the time required for one cycle of the sine wave to repeat. You will usually measure the period between consecutive positive zero-crossing points for the sine wave as shown in Figure 8-10, but you can measure the period between any two corresponding points on consecutive cycles. T for the sine wave in Figure 8-10 is five divisions.

Another characteristic of sine waves is f, the frequency. The frequency is the number of times per second that a sine wave repeats and is equal to the reciprocal of the period, $1/T$. The unit of frequency is hertz (Hz).

8.13 Reading Oscilloscope Displays

To understand oscilloscope measurements you must learn to read the display. Refer to Figure 8-11.

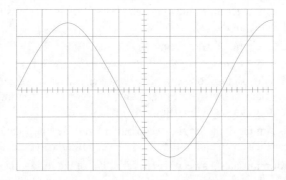

Figure 8-11: Example Oscilloscope Display

The oscilloscope display does not show numerical markings for the divisions. To determine the amplitude and period of the signal you must either

1) use the timebase and channel scale settings to convert the number of vertical and horizontal divisions into volts and seconds, or

2) use the cursors so that the Multisim program displays the measurement of interest in the display controls window.

For the waveform of Figure 8-11, determine and record the values of V_{pp}, V_p, T, and f for each of the timebase and channel settings in Table 8-1.

Table 8-1: Calculated Waveform Values for Figure 8-11

Waveform Value	Timebase = 5 µs/Div Channel = 200 µV/Div	Timebase = 200 µs/Div Channel = 50 mV/Div	Timebase = 1 ms/Div Channel = 10 V/Div
V_{pp}			
$V_p = V_{pp} / 2$			
T			
$f = 1/T$			

Conclusions for Part 1

Questions for Part 1

1) An oscilloscope shows the period of an ac waveform to be 3.5 horizontal divisions when the timebase control is set to "10 ms/Div". What is the measured period and frequency of the waveform?

2) You wish to view an ac waveform with a peak voltage V_p of 6.1 V. How many vertical divisions will the peak-to-peak signal be if the **Scale** setting is "5 V/Div"?

Part 2: Oscilloscope Measurements

8.14 Determining the Oscilloscope Timebase and Voltage Scale Settings

If you know the approximate frequency and amplitude of a signal that you wish to measure, it is a good idea for you to know the oscilloscope timebase and channel scale settings that you will use to measure the signal. This provides a check on the signal values that you expect to measure. Ideally, you would like one cycle to occupy the entire oscilloscope display so that the peak-to-peak signal amplitude is six divisions high, and the period is ten divisions wide. Practically, however, you must round up to the nearest 1-2-5 scale settings that will show the full signal amplitude and period.

For each of the signal amplitudes and frequencies listed in the headings of Table 8-2, calculate and record the ideal and practical (1-2-5 progression) oscilloscope timebase and channel scale settings.

Table 8-2: Calculated Oscilloscope Scale Settings

Waveform Value	$V_p = 200$ mV $f = 250$ kHz	$V_p = 10$ V $f = 400$ Hz	$V_p = 189$ V $f = 60$ kHz
$V_{pp} = 2 \times V_p$			
Ideal Channel Scale $= V_{pp} / 6$			
Practical Channel Scale			
$T = 1/f$			
Ideal Timebase Scale $= T / 10$			
Practical Channel Scale			

8.15 Multisim Verification

1) Open the Multisim file *DC_AC_Exp_08_Part_02a*.

2) For each of the timebase and channel settings in Table 8-1, use the right-click menu for the ac voltage source to set the **Voltage (Pk):** and **Frequency (F):** values to your calculated values of V_P and f in Table 8-1.

3) Simulate the circuit and verify that the oscilloscope display matches that of Figure 8-11 for your corresponding timebase and channel scale settings of Table 8-2.

4) Open the Multisim file *DC_AC_Exp_08_Part_02b*.

5) For each of the peak voltage and frequency settings in Table 8-2, use the right-click menu for the ac voltage source to set the **Voltage (Pk):** and **Frequency (F):** values to the values of V_P and f in Table 8-2.

6) Set the timebase and channel scale settings to the corresponding practical scale settings you calculated for Table 8-2 and simulate the circuit. Do your calculated settings allow the oscilloscope to display one complete cycle of the waveform? Can you decrease either the timebase or channel scale setting and still view the complete waveform?

Conclusions for Part 2

Questions for Part 2

1) What should the Multisim oscilloscope scale and timebase control settings be to view a 46 V_{pp} 60 Hz signal?

2) What is the maximum amplitude and minimum frequency of a signal that the Multisim oscilloscope can display with a scale setting of "20 V/Div" and timebase setting of "5 µs/Div"?

DC/AC Experiment 9 - Capacitors

9.1   Introduction

All circuit elements have three basic characteristics, although one is generally much larger so that the others can be ignored. You have already investigated resistance, the first characteristic. In this experiment you will examine the second characteristic, called *capacitance*. Capacitance C is defined as the ratio of the amount of charge, Q, on a device to the amount of voltage, V, required to store the charge (or, alternatively, to the voltage V that the stored charge Q develops across the device). The component whose primary characteristic is capacitance is called a *capacitor*. A capacitor with a high value of capacitance requires relatively little voltage to store a relatively large amount of charge. Conversely, a capacitor with a low capacitance value requires a relatively large voltage to store a relatively small amount of charge.

Capacitors exhibit certain important properties. One property is that voltage cannot change instantaneously across a capacitor. There is a lag between the time current through a capacitor changes and the time the voltage across the capacitor responds to that change. Another property is that the opposition of a capacitor to sinusoidal current, called *capacitive reactance*, is frequency-dependent. The capacitive reactance, X_C, is inversely proportional to the frequency, f, so that capacitive reactance increases as frequency decreases and vice-versa. The equation for the reactive capacitance X_C for a capacitance C is

$$X_C = 1 / (2\pi fC)$$

where X_C is in ohms, f is in hertz, and C is in farads.

In Part 1 of this experiment, you will examine how voltage changes across a capacitor by measuring an important property of resistive-capacitive (RC) circuits, called the time constant τ. The time constant is used when the capacitor is charging or discharging to a constant value. For an RC circuit,

$$\tau = RC$$

where τ is the time constant in seconds, C is the total amount of capacitance in farads, and R is the resistance in ohms through which the capacitance is charging or discharging. The voltage across a capacitor C will increase (or decrease) 63% of the way to its final value in one time constant.

In Part 2 of this experiment, you will measure and determine the relationship between capacitive reactance, sinusoidal voltage, and sinusoidal current in RC circuits.

9.2   Reading

Floyd and Buchla, *Electric Circuits Fundamentals, 8th Ed.*, Chapter 9

9.3   Key Objectives

Part 1:   Calculate the time constant for series RC circuits. Use the Multisim oscilloscope to experimentally verify the calculated values.

Part 2:   Calculate and experimentally verify the reactance, voltage, and current for series and parallel capacitive circuits.

9.4   Multisim Files

Part 1:   *DC_AC_Exp_09_Part_01a* and *DC_AC_Exp_09_Part_01b*

Part 2:   *DC_AC_Exp_09_Part_02a* and *DC_AC_Exp_09_Part_02b*

Part 1:   The *RC* Time Constant

9.5   Capacitors in Series

1)   Open the Multisim file *DC_AC_Exp_09_Part_01a*.

2)   Calculate to three significant digits the capacitor voltages $V_{C(1\tau)}$ and $V_{C(2\tau)}$ that correspond to 1 and 2 time constants of charging for the circuit.

$$V_{C(1\tau)} = 12 \text{ V } (1 - e^{-1}) = \textbf{7.59 V}$$

$$V_{C(2\tau)} = 12 \text{ V } (1 - e^{-2}) = \textbf{10.4 V}$$

3) For each of the SW_1 and SW_2 switch settings shown for the columns "Case 1", "Case 2", and "Case 3" in Table 9-1, calculate and record to three significant digits

    a. the total capacitance $C_{T(CALC)}$, and

    b. the time constant $\tau_{(CALC)}$.

4) Double-click on the oscilloscope to expand it.

5) Use the space bar to open SW_3.

6) Start the simulation.

7) Use the "1" and "2" keys to set SW_1 and SW_2 to the settings shown for the "Case 1" column in Table 9-1. Alternatively, you can click on the switches to open or close them.

8) Use the space bar to close SW_3.

9) When the trace is complete, stop the simulation.

10) Right-click on cursor (crosshair) 1 on the left side of the oscilloscope screen to open the cursor menu.

11) Click on **Select Trace ID**.

12) Click on down arrow (▼), select "Channel B" from the drop-down list, and click on the **OK** button.

13) Right-click on cursor 1 on the left side of the oscilloscope screen to open the cursor menu.

14) Click on **Set Y_Value =>**. This command will set cursor 1 at a specified voltage value on Channel B that first occurs to the right of its current position.

15) Enter your calculated value for $V_{C(1\tau)}$ and click the **OK** button. The cursor should now be at the specified voltage value on Channel B. To confirm this, verify that the measured voltage for **Channel B** in the **T1** row under the oscilloscope display matches the value that you entered.

16) Right-click on cursor 2 on the left side of the oscilloscope screen to open the cursor menu.

17) Click on **Set Y_Value <=**. This command will set cursor 2 at a specified voltage value on Channel B that first occurs to the left of its current position.

18) Enter your calculated value for $V_{C(2\tau)}$ and click the **OK** button. The cursor should now be at the specified voltage value on Channel B. To confirm this, verify that the measured voltage for **Channel B** in the **T2** row under the oscilloscope display matches the value that you entered.

19) Record to three significant digits the **Time** value in the **T2 – T1** row as "$\tau_{(MEAS)}$" in Table 9-1. This is the difference between the times for cursor 1 (1τ) and cursor 2 (2τ) and corresponds to the measured time constant for the circuit.

20) Repeat steps 5 through 19 for the "Case 2" and "Case 3" columns.

Table 9-1: Circuit Time Constants for Series Capacitors

Value	Case 1	Case 2	Case 3
SW_1	OPEN	CLOSED	OPEN
SW_2	CLOSED	OPEN	OPEN
$C_{T(CALC)}$			
$\tau_{(CALC)} = RC_{T(CALC)}$			
$\tau_{(MEAS)}$			

9.6   Capacitors in Parallel

1) Open the Multisim file *DC_AC_Exp_09_Part_01b*.

2) Calculate to three significant digits the capacitor voltages $V_{C(1\tau)}$ and $V_{C(2\tau)}$ that correspond to 1 and 2 time constants of charging for the circuit.

$$V_{C(1\tau)} =$$

$$V_{C(2\tau)} =$$

3) For each of the SW_1 and SW_2 switch settings shown for the columns "Case 1", "Case 2", and "Case 3" in Table 9-2, calculate and record to three significant digits

    a.   the total capacitance $C_{T(CALC)}$, and

    b.   the time constant $\tau_{(CALC)}$.

4) Double-click on the oscilloscope to expand it.

5) Use the space bar to open SW_3.

6) Start the simulation.

7) Use the "1" and "2" keys to set SW_1 and SW_2 to the indicated settings for the column. Alternatively, you can click on the switches to open or close them.

8) Use the space bar to close SW_3.

9) When the trace is complete, stop the simulation.

10) Right-click on cursor (crosshair) 1 on the left side of the oscilloscope screen to open the cursor menu.

11) Click on **Select Trace ID**.

12) Click on down arrow ($\blacktriangledown$), select "Channel B" from the drop-down list, and click the **OK** button.

13) Right-click on cursor 1 on the left side of the oscilloscope screen to open the cursor menu.

14) Click on **Set Y_Value =>**. This command will set cursor 1 at a specified voltage value on Channel B that first occurs to the right of its current position.

15) Enter your calculated value for $V_{C(1\tau)}$ and click the **OK** button. The cursor should now be at the specified voltage value on Channel B. To confirm this, verify that the measured voltage for **Channel B** in the **T1** row under the oscilloscope display matches the value that you entered.

16) Right-click on cursor 2 on the left side of the oscilloscope screen to open the cursor menu.

17) Click on **Set Y_Value <=**. This command will set cursor 2 at a specified voltage value on Channel B that first occurs to the left of its current position.

18) Enter your calculated value for $V_{C(2\tau)}$ and click the **OK** button. The cursor should now be at the specified voltage value on Channel B. To confirm this, verify that the measured voltage for **Channel B** in the **T2** row under the oscilloscope display matches the value that you entered.

19) Record to three significant digits the **Time** value in the **T2 – T1** row as "$\tau_{(MEAS)}$" in Table 9-2. This is the difference between the times for cursor 1 (1τ) and cursor 2 (2τ) and corresponds to the measured time constant for the circuit.

20) Repeat steps 5 through 19 for the "Case 2" and "Case 3" columns of Table 9-2.

Table 9-2: Circuit Time Constants for Parallel Capacitors

Value	Case 1	Case 2	Case 3
SW_1	CLOSED	OPEN	OPEN
SW_2	OPEN	CLOSED	OPEN
$C_{T(CALC)}$			
$\tau_{(CALC)} = RC_{T(CALC)}$			
$\tau_{(MEAS)}$			

Conclusions for Part 1

Questions for Part 1

1) Why didn't this experiment use the interval from when SW_1 closes to the time at which $V_{C(1\tau)}$ occurs to measure the time constant?

2) When measuring the time constants for the circuit in *DC_AC_Exp_09_Part_01b*, you find that $\tau_{(MEAS)} =$ 16.2 ms for Case 1. What would you suspect is the probable reason? If so, what would you expect $\tau_{(MEAS)}$ for Case 2 and Case 3 to be?

Part 2: Capacitors in AC Circuits

9.7 Capacitors in Series

1) Open the Multisim file *DC_AC_Exp_09_Part_02a*.
2) For each of the columns in Table 9-3, calculate and record to three significant digits:
   a. the capacitive reactance $X_{C1} = 1 / (2\pi f C_1)$ and $X_{C2} = 1/ (2\pi f C_2)$ for capacitors C_1 and C_2,
   b. the total capacitive reactance $X_{CT} = X_{C1} + X_{C2}$,
   c. the total rms current $I_{T(rms)} = V_S / X_{CT}$, and
   d. the rms voltages $V_{C1(rms)} = I_{T(rms)}X_{C1}$ and $V_{C2(rms)} = I_{T(rms)}X_{C2}$ for capacitors C_1 and C_2.

Table 9-3: Circuit Parameters for Series Capacitors

Value	Case 1	Case 2	Case 3	Value	Case 1	Case 2	Case 3
V_S	6.0 V	10.0 V	48.0 V	X_{CT}			
f	2.5 kHz	5 kHz	10 kHz	I_T			
X_{C1}				$V_{C1\,(rms)}$			
X_{C2}				$V_{C2\,(rms)}$			

3) Double-click on each of the multimeters to expand them.
4) Set XMM_1 for ac ammeter mode, and XMM_2 and XMM_3 for ac voltmeter mode.
5) Set the voltage and frequency of the ac source to the values indicated for the "Case 1" column in Table 9-3.
6) Start the simulation.
7) Measure and record to three significant digits the total ac current as "$I_{T(rms)}$", the ac voltage across C_1 as "$V_{C1(rms)}$", and the ac voltage across C_2 as the "$V_{C2(rms)}$" in the "Case 1" column of Table 9-4.
8) Stop the simulation.
9) Repeat steps 5 through 8 for the "Case 2" and "Case 3" columns of Table 9-4.

Table 9-4: Measured Circuit Values for Series Capacitors

Value	Case 1	Case 2	Case 3
$I_{T(rms)}$			
$V_{C1(rms)}$			
$V_{C2(rms)}$			

9.8 Capacitors in Parallel

1) Open the Multisim file *DC_AC_Exp_09_Part_02b*.

2) For each of the columns in Table 9-5, calculate and record to three significant digits:

    a. the capacitive reactances $X_{C1} = 1 / (2\pi f C_1)$ and $X_{C2} = 1 / (2\pi f C_2)$ for capacitors C_1 and C_2,

    b. the total capacitive reactance $X_{CT} = X_{C1} \parallel X_{C2}$,

    c. the total rms current $I_{T(rms)} = V_S / X_{CT}$, and

    d. the rms currents $I_{C1(rms)} = V_S / X_{C1}$ and $I_{C2(rms)} = V_S / X_{C2}$ for capacitors C_1 and C_2.

Table 9-5: Circuit Parameters for Parallel Capacitors

Value	Case 1	Case 2	Case 3	Value	Case 1	Case 2	Case 3
V_S	5.0 V	7.5 V	12.0 V	X_{CT}			
f	100 Hz	200 Hz	500 Hz	$I_{T(rms)}$			
X_{C1}				$I_{C1(rms)}$			
X_{C2}				$I_{C2(rms)}$			

3) Double-click on each of the multimeters to expand them.

4) Set each of the multimeters for ac ammeter mode.

5) Set the voltage and frequency of the ac source to the values indicated for the "Case 1" column in Table 9-5.

6) Start the simulation.

7) Measure and record to three significant digits the total ac current as "$I_{T(rms)}$", the ac voltage across C_1 as "$I_{C1(rms)}$", and the ac voltage across C_2 as "$I_{C2(rms)}$" in the "Case 1" column of Table 9-6.

8) Stop the simulation.

9) Repeat steps 5 through 8 for the "Case 2" and "Case 3" columns of Table 9-6.

Table 9-6: Measured Circuit Values for Parallel Capacitors

Value	Case 1	Case 2	Case 3
$I_{T(rms)}$			
$I_{C1(rms)}$			
$I_{C2(rms)}$			

Conclusions for Part 2

Questions for Part 2

1) In Table 9-3 and Table 9-5, what assumptions are made for calculating the value of X_{CT}?

2) From the calculated and measured voltages for series capacitors in Table 9-3 and Table 9-4, does Kirchhoff's voltage law appear to be valid for ac capacitor circuits? How does the data support your answer?

3) From the calculated and measured currents for parallel capacitors in Table 9-5 and Table 9-6, does Kirchhoff's current law appear to be valid for ac capacitor currents? How does the data support your answer?

DC/AC Experiment 10 - AC Response of *RC* Circuits

10.1 Introduction

Capacitors have a property, called reactance, that opposes sinusoidal current. Reactance differs from resistance in two ways. The first difference is that reactance varies with frequency. Capacitive reactance is inversely proportional to frequency, so that the reactance of a capacitor decreases as frequency increases and vice versa. A rule of thumb is that capacitors act like an open to dc and like a short to ac, so that capacitors block dc and pass ac. The second difference between reactance and resistance is that reactance introduces a 90° phase shift between voltage and current. Capacitive reactance has a –90° phase angle, so that the current through a capacitor **leads** the applied voltage by 90° (i.e., the current reaches its maximum value 90°, or one-quarter of a complete cycle, before the voltage reaches its maximum value). The combination of resistance and reactance, called *impedance*, of an *RC* circuit has a phase angle somewhere between 0° (purely resistive) and –90° (purely capacitive) so that the current will lead the voltage by somewhere between 0° and 90°.

RC circuits are called complex circuits because they contain both resistance and reactance that combine to create impedance. The subject of complex numbers and arithmetic is beyond the scope of this manual, but excellent discussions can be found on the Internet and in electronic texts such as Floyd's *Principles of Electric Circuits, 9th Edition*. For series *RC* circuits, the impedance magnitude, Z_T, and phase, φ, be found from

$$Z_T = \sqrt{R^2 + X_C{}^2} \text{ and } \varphi = -\tan^{-1}\frac{X_C}{R}$$

In Part 1 of this experiment, you will calculate the voltages for series *RC* circuits and use Multisim to verify your answers. In Part 2, you will use an oscilloscope to measure amplitude and phase and determine the frequency response of *RC* low-pass and high-pass filters.

10.2 Reading

Floyd and Buchla, *Electric Circuits Fundamentals, 8th Ed.*, Chapter 10

10.3 Key Objectives

Part 1: Calculate the amplitude and phase of *RC* circuits and use Multisim to verify your results.

Part 2: Measure the output amplitude and phase to plot the frequency response of low-pass and high-pass *RC* circuits.

10.4 Multisim Files

Part 1: *DC_AC_Exp_10_Part_01*

Part 2: *DC_AC_Exp_10_Part_02a* and *DC_AC_Exp_10_Part_02b*

Part 1: Amplitude and Phase of *RC* Circuits

1) For each column in Table 10-1:

   a. Use f and C to calculate and record the capacitive reactance $X_C = 1 / (2\pi f C)$.

   b. Use R and X_C to calculate and record the impedance magnitude $Z_T = \sqrt{R^2 + X_C{}^2}$.

   c. Use V_S, R, and Z_T to calculate and record the resistor voltage magnitude $V_R = V_S (R / Z_T)$.

   d. Use R and X_C to calculate and record the resistor voltage phase $\varphi_R = \tan^{-1}(X_C / R)$.

   e. Use V_S, X_C, and Z_T to calculate and record the capacitor voltage magnitude $V_C = V_S (X_C / Z_T)$.

   f. Use R and X_C to calculate and record the capacitor voltage phase $\varphi_C = \tan^{-1}(X_C / R) - 90°$.

Table 10-1: Calculated *RC* Circuit Values

Value	*f* = 1 kHz	*f* = 5 kHz	*f* = 12.5 kHz	Value	*f* = 1 kHz	*f* = 5 kHz	*f* = 12.5 kHz
V_S	1.0 $V_p \angle 0°$	2.5 $V_p \angle 0°$	0.5 $V_p \angle 0°$	V_R			
R	100 Ω	75 kΩ	4.3 kΩ	φ_R			
C	1.0 µF	220 pF	4.7 nF	V_C			
X_C				φ_C			
Z_T							

2) Open the Multisim file *DC_AC_Exp_10_Part_01*.

3) Change the values of the voltage source V_S amplitude, phase, and frequency, resistor R, and capacitor C to match those for the "*f* = 1 kHz" column in Table 10-2.

4) Use *f* to calculate and record the period $T = 1 / f$ in Table 10-2.

5) Run the simulation to capture at least one complete waveform for Channel A. Adjust the timebase and channel controls as needed to view the entire waveform.

6) Set cursor 2 to the first maximum of Channel A.

7) Set cursor 1 to the maximum value of Channel B that is closest to cursor 2.

8) Record the **Channel B** value in the **T2** row under the oscilloscope display as V_R in the first column of Table 10-2.

9) Calculate the phase for the resistor voltage as follows:

    a. Record the **Time** value in the **T2 – T1** row under the oscilloscope display as Δt_R in Table 10-2.

    b. Calculate and record the resistor voltage phase $\varphi_R = (\Delta t_R / T) \times 360°$ in Table 10-2.

10) Swap the positions of *R* and *C* in the circuit so that Channel B measures the voltage across C_1.

11) Run the simulation to capture at least one complete waveform for Channel A. Adjust the timebase and channel controls as needed to view the entire waveform.

12) Set cursor 2 to the first maximum of Channel A.

13) Set cursor 1 to the maximum value of Channel B that is closest to cursor 2.

14) Record the **Channel B** value in the **T2** row under the oscilloscope display as V_C for the appropriate column of Table 10-2.

15) Calculate the phase for the capacitor voltage as follows:

    a. Record the **Time** value in the **T2 – T1** row under the oscilloscope display as Δt_C in Table 10-2.

    b. Calculate and record the capacitor voltage phase $\varphi_C = [(\Delta t_C / T) \times 360°] - 90°$ in Table 10-2.

16) Repeat steps 3 through 16 for the "*f* = 5 kHz" and "*f* = 12.5 kHz" columns.

Table 10-2: Measured *RC* Circuit Values

Value	*f* = 1 kHz	*f* = 5 kHz	*f* = 12.5 kHz	Value	*f* = 1 kHz	*f* = 5 kHz	*f* = 12.5 kHz
V_S	1.0 $V_p \angle 0°$	2.5 $V_p \angle 0°$	0.5 $V_p \angle 0°$	V_R			
R	100 Ω	75 kΩ	4.3 kΩ	φ_R			
C	1.0 µF	220 pF	4.7 nF	Δt_C			
T				V_C			
Δt_R				φ_C			

Observations:

Conclusions for Part 1

Questions for Part 1

1) From the data in Table 10-1, how does the value of X_C / R_1 affect the phase angle?

2) Explain why the equation $\varphi = [\Delta t / T] \times 360°$ gives the phase difference between two waveforms.

Part 2: Frequency Response of *RC* Circuits

10.5 High-Pass RC Filter

1) Open the Multisim file *DC_AC_Exp_10_Part_02a*.
2) For each of the frequencies in Table 10-3:
   - a. Use f and C to calculate and record the capacitive reactance $X_C = 1 / (2\pi fC)$.
   - b. Use R and X_C to calculate and record the total impedance $Z_T = \sqrt{R^2 + X_C^2}$.
   - c. Use V_S, R, and Z_T to calculate and record the output voltage magnitude $V_{out} = V_S (R / Z_T)$.
   - d. Use R and X_C to calculate and record the output voltage phase angle $\varphi_{out} = \tan^{-1}(X_C / R)$.

Table 10-3: Calculated Values for High-Pass Filter Circuit ($R = 10$ kΩ, $C = 80$ nF)

f (Hz)	X_C	Z_T	V_{out}	φ_{out}	f (Hz)	X_C	Z_T	V_{out}	φ_{out}
10					500				
20					1000				
50					2000				
100					5000				
200					10,000				

3) For each of the frequencies in Table 10-4:
   - a. Set the voltage source to the indicated frequency. For frequencies of 2 kHz and above, change the time step for the oscilloscope sampling as follows:
     - i. From the **Simulate** menu, select **Interactive Simulation Settings...**.
     - ii. Select the **Maximum time step (TMAX)** radio button.

      iii.   Enter "1e-007" in the **TMAX** text box so that the oscilloscope samples are no further than 0.1 μs apart.

   b.   Start the simulation. Allow the simulation to run until the Channel B waveform settles to a stable value.

   c.   Adjust the timebase and channel controls as needed to view one complete waveform with between 2 and 3 vertical divisions of amplitude for each channel.

   d.   Stop the simulation.

   e.   Measure and record the peak amplitude on Channel B as V_{out}.

   f.   Measure and record the phase angle between Channel B and Channel A as φ_{out}.

4) Calculate and record the common (Base 10) logarithm of each measured voltage value in Table 10-4.

5) Calculate and record the common logarithm of each frequency in Table 10-4.

Table 10-4: Measured Data for High-Pass RC Filter

f (Hz)	V_{out}	φ_{out}	$Log_{10} V$	$Log_{10} f$	f (Hz)	V_{out}	φ_{out}	$Log_{10} V$	$Log_{10} f$
10					500				
20					1000				
50					2000				
100					5000				
200					10,000				

6) Use your data from Table 10-4 to plot "$Log_{10} V$" vs. "$Log_{10} f$" on Plot 10-1.

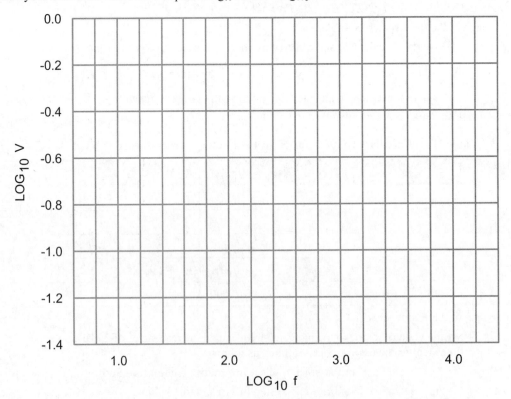

Plot 10-1: High-Pass RC Filter Voltage vs. Frequency Log Plot

7) Use your data from Table 10-4 to plot the phase angle "φ_{out}" vs. "$\text{Log}_{10}\,f$" on Plot 10-2.

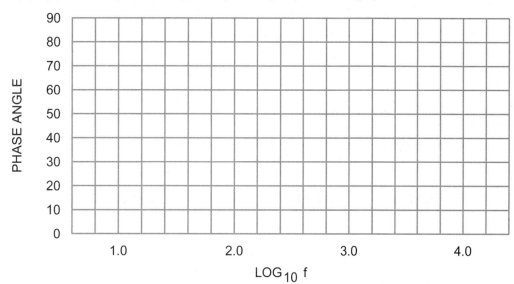

Plot 10-2: High-Pass *RC* Filter Phase vs. Frequency Log Plot

10.6 Low-Pass *RC* Filter

1) Open the Multisim file *DC_AC_Exp_10_Part_02b*.

2) For each of the frequencies in Table 10-5:

    a. Use f and C to calculate and record the capacitive reactance $X_C = 1 / (2\pi f C)$.

    b. Use R and X_C to calculate and record the total impedance $Z_T = \sqrt{R^2 + X_C^2}$.

    c. Use V_S, X_C, and Z_T to calculate and record the output voltage magnitude $V_{out} = V_S\,(X_C / Z_T)$.

    d. Use X_C and Z_T to calculate and record the output voltage phase angle $\varphi_{out} = \tan^{-1}(X_C / Z_T) - 90°$.

Table 10-5: Calculated Values for Low-Pass *RC* Filter Circuit (R = 2.4 kΩ, C = 130 nF)

f (Hz)	X_C	Z_T	V_{out}	φ_{out}	f (Hz)	X_C	Z_T	V_{out}	φ_{out}
10					500				
20					1000				
50					2000				
100					5000				
200					10,000				

3) For each of the value in Table 10-6:

    a. Set the voltage source to the indicated frequency. For frequencies of 2 kHz and above change the time step for the oscilloscope sampling as follows:

        i. From the **Simulate** menu, select **Interactive Simulation Settings....**

        ii. Select the **Maximum time step (TMAX)** radio button.

        iii. Enter "1e-007" in the **TMAX** text box so that the oscilloscope samples are no further than 0.1 μs apart.

    b. Start the simulation. Allow the simulation to run until the Channel B waveform settles to a stable value so that the positive and negative peaks are vertically centered.

  c. Adjust the timebase and channel controls as needed to view one complete waveform with between 2 and 3 vertical divisions of amplitude for each channel.

  d. Stop the simulation.

  e. Measure the amplitude on Channel B. Record the value as V_{out}.

  f. Measure the phase between Channel B and Channel A. Record this value as φ_{out}.

4) Calculate the common (base 10) logarithm of each measured voltage value in Table 10-6. Record these values as "$Log_{10} V$".

5) Calculate the common logarithm of each frequency in Table 10-6. Record these values as "$Log_{10} f$".

Table 10-6: Measured Data for Low-Pass *RC* Filter

f (Hz)	V_{out}	φ_{out}	$Log_{10} V$	$Log_{10} f$	f (Hz)	V_{out}	φ_{out}	$Log_{10} V$	$Log_{10} f$
10					500				
20					1000				
50					2000				
100					5000				
200					10,000				

6) Use your data from Table 10-6 to plot "$Log_{10} V$" vs. "$Log_{10} f$" on Plot 10-3.

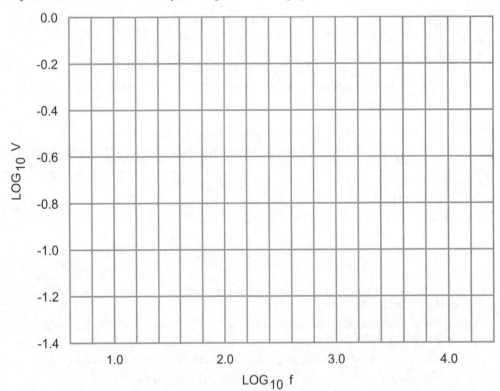

Plot 10-3: Low-Pass *RC* Filter Voltage vs. Frequency Log Plot

7) Use your data from Table 10-6 to plot "φ_{out}" vs. "$\text{Log}_{10} f$" on Plot 10-4.

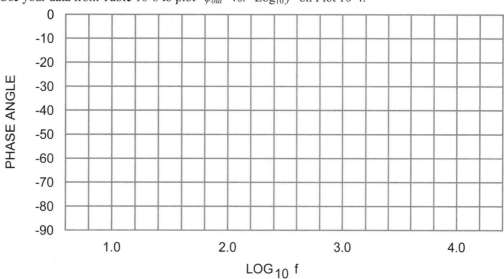

Plot 10-4: Low-Pass *RC* Filter Phase vs. Frequency Log Plot

Conclusions for Part 2

Questions for Part 2

1) As the frequency increases, what is the approximate frequency at which the output of the high-pass filter begins to increase rapidly?

2) As the frequency increases, what is the approximate frequency at which the output of the low-pass filter begins to decrease rapidly?

3) Why was it necessary to change the simulation settings for frequencies of 2 kHz and higher?

DC/AC Experiment 11 - Inductors

11.1 Introduction

In previous experiments you examined resistance and capacitance, two of the three basic characteristics of electric circuits. In this experiment you will examine the third characteristic, called *inductance*. Inductance L is defined as the ratio of the amount of voltage, V, on a device to the rate at which current is changing, $\Delta I / \Delta t$, required to induce the voltage. The component whose primary property is inductance is called an *inductor*. An inductor with a high inductance value requires current to change relatively slowly to induce a relatively large voltage. Conversely, an inductor with a low inductance value requires current to change relatively quickly to induce the same voltage.

Inductors have certain important properties. One property is that current cannot change instantaneously through an inductor. There is a lag between the time voltage across an inductor changes and the time the current through the inductor responds to that change. Another property is that the opposition of an inductor to sinusoidal current, called *inductive reactance*, is frequency-dependent. The inductive reactance X_L is directly proportional to the frequency f, so that inductive reactance increases as frequency increases and decreases as frequency decreases. The equation for the inductive reactance X_L for an inductance L is

$$X_L = 2\pi f L$$

where X_L is in ohms, f is in hertz, and L is in henries.

In Part 1 of this experiment you will examine how voltage changes across an inductor by measuring an important property of resistive-inductive (*RL*) circuits, called the time constant τ. For an *RL* circuit, $\tau = L / R$, where τ is the time constant in seconds, L is the total amount of inductance in henries, and R is the resistance in ohms through which the inductor is energizing or de-energizing. The voltage across an inductor L will increase (or decrease) 63% of the way to its final value in one time constant.

In Part 2 of this experiment you will measure and determine the relationship between inductive reactance, voltage, and current in *RL* circuits.

11.2 Reading

Floyd and Buchla, *Electric Circuits Fundamentals, 8th Ed.*, Chapter 11

11.3 Key Objectives

Part 1: Calculate the time constant for series *RL* circuits. Use the Multisim oscilloscope to experimentally verify the calculated values.

Part 2: Calculate and experimentally verify the reactance, voltage, and current for series and parallel inductive circuits.

11.4 Multisim Files

Part 1: *DC_AC_Exp_11_Part_01a* and *DC_AC_Exp_11_Part_01b*

Part 2: *DC_AC_Exp_11_Part_02a* and *DC_AC_Exp_11_Part_02b*

Part 1: The *RL* Time Constant

11.5 Inductors in Series

1) Open the Multisim file *DC_AC_Exp_11_Part_01a*.

2) Calculate to three significant digits the inductor voltages $V_{L(1\tau)}$ and $V_{L(2\tau)}$ that correspond to 1 and 2 time constants of de-energizing for the circuit.

$$V_{L(1\tau)} = 5 \text{ V } (e^{-1}) =$$

$$V_{L(2\tau)} = 5 \text{ V } (e^{-2}) =$$

3) For each of the SW_1 and SW_2 switch settings shown for the columns "Case 1", "Case 2", and "Case 3" in Table 11-1, calculate and record to three significant digits:

    a.   the total inductance $L_{T(CALC)}$, and

    b.   the time constant $\tau_{(CALC)}$.

4) Double-click on the oscilloscope to expand it.

5) Start the simulation.

6) Use the "1" and "2" keys to set SW_1 and SW_2 to the settings shown for the "Case 1" column in Table 11-1. Alternatively, you can click on the switches to open or close them.

7) When the trace is complete, stop the simulation.

8) Position cursor 1 to your calculated value for $V_{L(1\tau)}$ on Channel B. Verify that the measured voltage for **Channel B** in the **T1** row under the oscilloscope display matches the value that you entered.

9) Position cursor 2 to your calculated value for $V_{L(2\tau)}$ on Channel B. Verify that the measured voltage for **Channel B** in the **T1** row under the oscilloscope display matches the value that you entered.

10) Record to three significant digits the **Channel B** value in the **T2 – T1** row as "$\tau_{(MEAS)}$" in Table 11-1. This is the difference between the times for cursor 1 (1τ) and cursor 2 (2τ) and corresponds to the measured time constant for the circuit.

11) Repeat steps 5 through 10 for the "Case 2" and "Case 3" columns.

Table 11-1: Circuit Time Constants for Series Inductors

Value	Case 1	Case 2	Case 3
SW_1	OPEN	CLOSED	OPEN
SW_2	CLOSED	OPEN	OPEN
$L_{T(CALC)}$			
$\tau_{(CALC)} = L_{T(CALC)} / R$			
$\tau_{(MEAS)}$			

11.6 Inductors in Parallel

1) Open the Multisim file *DC_AC_Exp_11_Part_01b*.

2) Calculate to three significant digits the inductor voltages $V_{L(1\tau)}$ and $V_{L(2\tau)}$ for the circuit that correspond to 1 and 2 time constants of de-energizing.

$$V_{L(1\tau)} = 10 \text{ V } (e^{-1}) =$$
$$V_{L(2\tau)} = 10 \text{ V } (e^{-2}) =$$

3) For each of the SW_1 and SW_2 switch settings shown for the columns "Case 1", "Case 2", and "Case 3" in Table 11-2, calculate and record to three significant digits

    a.   the total inductance $L_{T(CALC)}$, and

    b.   the time constant $\tau_{(CALC)}$.

4) Double-click on the oscilloscope to expand it.

5) Start the simulation.

6) Use the "1" and "2" keys to set SW_1 and SW_2 to the indicated settings for the column. Alternatively, you can click on the switches to open or close them.

7) When the trace is complete, stop the simulation.

8) Position cursor 1 to your calculated value for $V_{L(1\tau)}$ on Channel B. Verify that the measured voltage for **Channel B** in the **T1** row under the oscilloscope display matches the value that you entered.

9) Position cursor 2 to your calculated value for $V_{L(2\tau)}$ on Channel B. Verify that the measured voltage for **Channel B** in the **T1** row under the oscilloscope display matches the value that you entered.

10) Record to three significant digits the **Channel B** value in the **T2 – T1** row as "$\tau_{(MEAS)}$" in Table 11-2. This is the difference between the times for cursor 1 (1τ) and cursor 2 (2τ) and corresponds to the measured time constant for the circuit.

11) Repeat steps 5 through 10 for the "Case 2" and "Case 3" columns of Table 11-2.

Table 11-2: Circuit Time Constants for Parallel Inductors

Value	Case 1	Case 2	Case 3
SW_1	CLOSED	OPEN	OPEN
SW_2	OPEN	CLOSED	OPEN
$L_{T(CALC)}$			
$\tau_{(CALC)} = L_{T(CALC)} / R$			
$\tau_{(MEAS)}$			

Conclusions for Part 1

Questions for Part 1

1) For the circuit in *DC_AC_Exp_11_Exp_01a*, how does the voltage waveform across the inductor when the V_1 changes from 0 V to 5 V differ from when V_1 changes from 5 V to 0 V?

2) For the circuit in *DC_AC_Exp_11_Exp_01b*, why does the time constant decrease when both inductors are switched into the circuit?

Part 2: Inductors in AC Circuits

11.7 Inductors in Series

1) Open the Multisim file *DC_AC_Exp_11_Part_02a*.

2) For each of the columns in Table 11-3, calculate and record to three significant digits:
   a. the inductive reactance $X_{L1} = 2\pi fL_1$ and $X_{L2} = 2\pi fL_2$ for inductors L_1 and L_2,
   b. the total inductive reactance $X_{LT} = X_{L1} + X_{L2}$,
   c. the total rms current $I_{T(rms)} = V_S / X_{LT}$, and
   d. the rms voltages $V_{L1(rms)} = I_{T(rms)}X_{L1}$ and $V_{L2(rms)} = I_{T(rms)}X_{L2}$ for inductors L_1 and L_2.

Table 11-3: Circuit Parameters for Series Inductors

Value	Case 1	Case 2	Case 3	Value	Case 1	Case 2	Case 3
V_S	10.0 V	5.0 V	1.25 V	X_{LT}			
f	100 Hz	200 Hz	500 Hz	$I_{T(rms)}$			
X_{L1}				$V_{L1(rms)}$			
X_{L2}				$V_{L2(rms)}$			

3) Double-click on each of the multimeters to expand them.

4) Set XMM_1 for ac ammeter mode, and XMM_2 and XMM_3 for ac voltmeter mode.

5) Set the voltage and frequency of the ac source to the values indicated for the "Case 1" column in Table 11-3.

6) Start the simulation.

7) Measure and record to three significant digits the total ac current as the "$I_{T(rms)}$", the ac voltage across L_1 as the "$V_{L1(rms)}$", and the ac voltage across L_2 as the "$V_{L2(rms)}$" for the "Case 1" column of Table 11-4.

8) Stop the simulation.

9) Repeat steps 5 through 8 for the "Case 2" and "Case 3" columns of Table 11-4.

Table 11-4: Measured Values for Series Inductors

Value	Case 1	Case 2	Case 3
$I_{T(rms)}$			
$V_{L1(rms)}$			
$V_{L2(rms)}$			

11.8 Inductors in Parallel

1) Open the Multisim file *DC_AC_Exp_11_Part_02b*.

2) For each of the columns in Table 11-5, calculate and record to three significant digits:

    a. the inductive reactance $X_{L1} = 2\pi f L_1$ and $X_{L2} = 2\pi f L_2$ for inductors L_1 and L_2,

    b. the total inductive reactance $X_{LT} = X_{L1} \parallel X_{L2}$,

    c. the total rms current $I_{T(rms)} = V_S / X_{LT}$, and

    d. the rms currents $I_{L1(rms)} = V_S / X_{L1}$ and $I_{L2(rms)} = V_S / X_{L2}$ for inductors L_1 and L_2.

Table 11-5: Circuit Parameters for Parallel Inductors

Value	Case 1	Case 2	Case 3	Value	Case 1	Case 2	Case 3
V_S	3.3 V	10.0 V	15.0 V	X_{LT}			
f	250 Hz	750 Hz	2.25 kHz	$I_{T(rms)}$			
X_{L1}				$I_{L1(rms)}$			
X_{L2}				$I_{L2(rms)}$			

3) Double-click on each of the multimeters to expand them.

4) Set each of the multimeters for ac ammeter mode.

5) Set the voltage and frequency of the ac source to the values indicated for the "Case 1" column in Table 11-5.

6) Start the simulation.

7) Measure and record to three significant digits the total ac current as the "$I_{T(rms)}$", the ac current through L_1 as the "$I_{L1(rms)}$", and the ac current through L_2 as the "$I_{L2(rms)}$" for the "Case 1" column of Table 11-6.

8) Stop the simulation.

9) Repeat steps 5 through 8 for the "Case 2" and "Case 3" columns of Table 11-6.

Table 11-6: Measured Values for Parallel Inductors

Value	Case 1	Case 2	Case 3
Measured $I_{T(rms)}$			
Measured $I_{L1(rms)}$			
Measured $I_{L2(rms)}$			

Conclusions for Part 2

Questions for Part 2

1) From the data in Table 11-3, does the voltage divider rule appear to be valid for series ac inductor circuits? How does the data support your answer?

2) From the calculated and measured voltages for series inductors in Table 11-3 and Table 11-4, does Kirchhoff's voltage law appear to be valid for ac inductor circuits? How does the data support your answer?

3) From the calculated and measured currents for parallel inductors in Table 11-5 and Table 11-6, does Kirchhoff's current law appear to be valid for ac inductor circuits? How does the data support your answer?

DC/AC Experiment 12 - AC Response of *RL* Circuits

12.1 Introduction

As with capacitors, inductors have reactance that opposes sinusoidal current. Inductive reactance is directly proportional to frequency, so that the reactance of an inductor decreases as frequency decreases and increases as frequency increases. A rule of thumb is that inductors act like an short to dc and like an open to ac, so that inductors pass dc and block ac. Inductive reactance differs from capacitive reactance in that inductive reactance has a +90° phase angle, so that the current through an inductor **lags** the applied voltage by 90° (i.e., the current reaches its maximum value 90°, or one-quarter of a complete cycle, after the voltage reaches its maximum value). The impedance of an *RL* circuit will have a phase angle somewhere between 0° (purely resistive) and +90° (purely inductive) so that the current will lead the voltage by somewhere between 0° and 90°. A way to remember the difference in phase between voltage and current for inductive and capacitive circuits is the phrase "ELI the ICE man". ELI indicates that voltage (*E*) leads current (*I*) for inductance (*L*), and ICE indicates that current (*I*) leads voltage (*E*) for capacitance (*C*).

As with *RC* circuits, *RL* circuits contain both resistance and reactance. For series *RL* circuits, the impedance magnitude Z_T and phase φ can be found from

$$Z_T = \sqrt{R^2 + X_L^{\,2}} \ \text{ and } \ \varphi = \tan^{-1}\frac{X_L}{R}$$

In Experiment 10 you used an oscilloscope to measure the amplitude and phase of signals in *RC* circuits, which is the usual method for analyzing practical complex circuits. In this experiment, you will use the Bode plotter. The Bode plotter is not a real instrument, although combinations of other test equipment can emulate its functions. The Bode plotter creates plots similar to the log plots you created for the *RC* high-pass and low-pass filters, although it measures amplitude in specific logarithmic units called decibels (dB). The advantage of Bode plots is that it linearizes the amplitude and phase angle so that you can quickly visualize and estimate the frequency response of circuits. Figure 12-1 shows the idealized amplitude and phase angle Bode plots for the *RL* low-pass filter. Note that the frequency axis is marked in decades, or so that each division indicates an increase or decrease in frequency by a factor of ten.

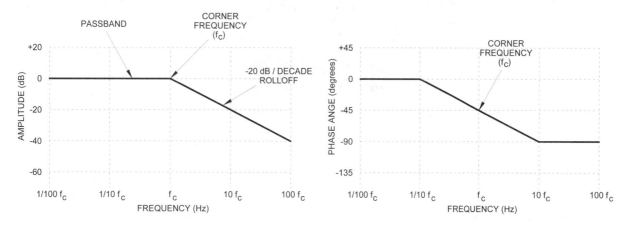

Figure 12-1: Idealized Bode Plot for *RL* Low-Pass Filter

In Part 1 of this experiment, you will calculate the voltages for series *RL* circuits and use Multisim to verify your answers. In Part 2, you will measure amplitude and phase to determine the frequency response of *RL* low-pass and high-pass filters.

12.2 Reading

Floyd and Buchla, *Electric Circuits Fundamentals, 8th Ed.*, Chapter 12

12.3 Key Objectives

Part 1:   Calculate the amplitude and phase of *RL* circuits and use the Multisim Bode plotter to verify your results.

Part 2:   Use the Bode plotter to measure the output amplitude and phase to determine the frequency response of low-pass and high-pass *RL* circuits and sketch the idealized amplitude and phase Bode plots of the filters.

12.4 Multisim Files

Part 1:   DC_AC_Exp_12_Part_01

Part 2:   *DC_AC_Exp_12_Part_02a* and *DC_AC_Exp_12_Part_02b*

Part 1:   Amplitude and Phase of *RL* Circuits

12.5 The Multisim Bode Plotter

The Multisim Bode plotter offers a convenient way to determine a circuit's frequency response. Figure 12-2 shows the **Bode Plotter** tool on the **Instruments Toolbar**. Figure 12-3 shows the minimized and expanded views of the Bode plotter.

Figure 12-2: The Bode Plotter Tool

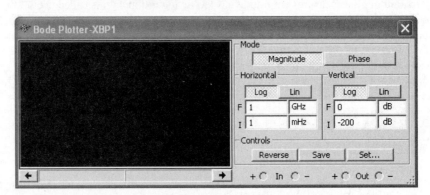

Figure 12-3: Minimized and Expanded Views of the Bode Plotter

The Bode plotter has five sections of settings.

12.5.1 Bode Plotter Graphical Display

The Bode plotter graphical display is the large area on the left side of the instrument. The display, shown in Figure 12-4, is similar to an oscilloscope display but shows frequency rather than time on the horizontal axis.

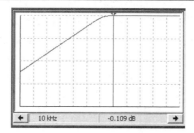

Figure 12-4: Bode Plotter Graphical Display

The Bode plotter magnitude and phase displays each possess one cursor so that you can select specific points on the Bode plot. The status bar below the display provides the magnitude (or phase) of the selected point. Note that the left and right arrows move the cursor and do not scroll the graphical display. The Bode plotter always uses the horizontal and vertical settings that you specify to fit the data into a single screen. You can also use the mouse to drag the cursor or right-click the cursor to open a right-click menu that is similar to that for the oscilloscope cursor.

12.5.2 Mode settings

These settings determine the Bode plotter mode of operation. **Magnitude** will display the magnitude (gain) of the circuit frequency response. **Phase** will display the phase of the circuit frequency response.

12.5.3 Horizontal settings

These settings configure the horizontal axis for the Bode plotter display. **Log** specifies a logarithmic frequency scale and **Lin** specifies a linear frequency scale. The **F** value sets the final frequency and the **I** value sets the initial frequency for the measured frequency range.

12.5.4 Vertical settings

These settings configure the vertical axis for the Bode plotter display. **Log** specifies a logarithmic scale and **Lin** specifies a linear scale. The **F** value sets the final vertical axis value and the **I** value sets the initial vertical axis value.

If you select **Magnitude** as the Bode plotter mode then the vertical axis will represent the gain (that is, the ratio of V_{out} to V_{in}). A linear scale will represent the magnitude as the ratio of V_{out} / V_{in}. A logarithmic scale will represent the magnitude in decibels (dB), which for voltage is 20 log (V_{out} / V_{in}).

If you select **Phase** as the Bode plotter mode, then the vertical axis will be linear and have units of degrees.

12.5.5 Plotter controls

These settings allow you to work with the plotter data.

Reverse allows you to select between a white or black background for the Bode plotter display, just as for the two-channel oscilloscope display.

Save allows you to save the measured data to either a .BOD text file or a .TDM binary file.

Set enables you to determine the resolution of the displayed data (i.e., the total number of data points that the Bode plotter will collect).

12.6 Using the Bode Plotter

1) For each column in Table 12-1:

    a. Use f and L to calculate and record the inductive reactance $X_L = 2\pi f L$.

    b. Use R and X_L to calculate and record the impedance magnitude $Z_T = \sqrt{R^2 + X_L^2}$.

    c. Use V_S, R, and Z_T calculate and record the resistor voltage magnitude $V_R = V_S (R / Z_T)$.

    d. Use R and X_L to calculate and record the phase $\varphi_R = -\tan^{-1} (X_L / R)$.

    e. Use V_S, X_L, and Z_T to calculate and record the inductor voltage magnitude $V_L = V_S (X_L / Z_T)$.

    f. Use R and X_L to calculate and record the phase $\varphi_L = 90° - \tan^{-1} (X_L / R)$.

    g. Use V_R to calculate and record V_R (dB) = 20 log (V_R / V_S).

    h. Use V_L to calculate and record V_L (dB) = 20 log (V_L / V_S).

Table 12-1: Calculated *RL* Circuit Values

Value	Case 1	Case 2	Case 3	Value	Case 1	Case 2	Case 3
V_S	2.5 V$_p$ $\angle$ 0°	5.0 V$_p$ $\angle$ 0°	100 mV$_p$ $\angle$ 0°	V_R			
f	500 Hz	2.5 kHz	20 kHz	φ_R			
R	100 Ω	7.5 kΩ	4.3 kΩ	V_L			
L	62 mH	330 mH	33 mH	φ_L			
X_L				V_R (dB)			
Z_T				V_L (dB)			

2) Open the Multisim file *DC_AC_Exp_12_Part_01*.

3) Select the **Bode Plotter** tool from the **Instruments Toolbar** and place it above the circuit.

4) Connect the Bode plotter input across V_{S1} by connecting IN "+" between V_1 and L and IN "–" to ground. The Bode plotter will use V_S as the reference for the magnitude and phase measurements. As a rule you will always connect the Bode plotter input directly to the source so that the reference signal does not change with frequency.

5) Connect the Bode plotter output across R by connecting OUT "+" between L and R and OUT "–" to ground. The Bode plotter will measure the magnitude and phase across R relative to the input voltage (V_S).

6) Double-click the Bode plotter to open the enlarged view.

7) Change the values of the voltage source amplitude, phase, and frequency, resistor, and inductor to match those for the "Case 1" column in Table 12-1.

8) Change the final (maximum) frequency to 100 times the frequency in Table 12-1 by entering that value in the **F** text box in the **Horizontal** control section.

9) Change the initial (minimum) frequency to 1/100th the frequency in Table 12-1 by entering that value in the **I** text box in the **Horizontal** control section.

10) Start the simulation and wait for the Bode plotter to display the plot.

11) Stop the simulation. The display should look similar to Figure 12-5. As you can see, the magnitude of the output decreases as the frequency increases and L appears increasingly like an open. Below a certain frequency, called the **corner frequency**, the magnitude levels out. The corner frequency occurs at the frequency at which the resistance and reactance in a circuit are equal.

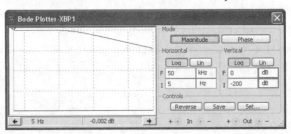

Figure 12-5: Example of Bode Plot Magnitude Display

12) Set the **Mode** setting to **Magnitude**.

13) Right-click the magnitude cursor to open the right-click menu and select **Set X_Value**.

14) Enter the frequency value for "Case 1" in Table 12-1 to reposition the cursor. The frequency and magnitude of the resistor voltage in decibels appear below the display, as shown in Figure 12-6. Record the magnitude to three significant digits as V_R (dB) in the "Case 1" column of Table 12-2.

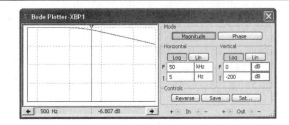

Figure 12-6: Bode Plotter with Repositioned Magnitude Cursor

15) Change the **Mode** setting to **Phase**.

16) Change the final (maximum) phase value to +90° by entering "0" in the **F** text box in the **Vertical** control section.

17) Change the initial (minimum) phase value to −90° by entering "−90" in the **I** text box in the **Vertical** control section.

18) Right-click the phase cursor to open the right-click menu and select **Set X_Value**.

19) Enter the frequency value for Case 1 shown in Table 12-1 to reposition the cursor. The frequency and phase of the resistor voltage in degrees appear below the display, as shown in Figure 12-7. Record the phase to three significant digits as φ_R in the "Case 1" column of Table 12-2.

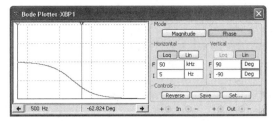

Figure 12-7: Example of Bode Plot Phase Display

20) Connect the Bode plotter output across L by connecting OUT "+" between V_S and L and OUT "−" between L and R. Summarize how the magnitude and phase values differ from those of the resistor.

21) Measure the magnitude V_L (dB) and phase φ_L for the inductor at the frequency for Case 1. Record these values in Table 12-2.

22) Repeat steps 7 through 21 for the "Case 2" and "Case 3" circuit values.

Table 12-2: Measured *RL* Circuit Values

Value	Case 1	Case 2	Case 3
V_R (dB)			
φ_R			
V_L (dB)			
φ_L			

Observations:

Conclusions for Part 1

Questions for Part 1

1) From the magnitude waveform across R, what type of filter would the circuit in *DC_AC_Exp_12_Part_01* be if the output was taken from across the resistor?

2) How does the magnitude waveform of the resistor voltage differ from that of the inductor voltage? Explain your answer.

Part 2: Frequency Response of *RL* Circuits

12.7 Low-Pass *RL* Filter

1) Open the Multisim file *DC_AC_Exp_12_Part_02a*.

2) For each of the frequencies in Table 12-3:

    a. Use f and L to calculate and record the inductive reactance $X_L = 2\pi f L$.

    b. Use R and X_L to calculate and record the impedance magnitude $Z_T = \sqrt{R^2 + X_L^2}$.

    c. Use V_S, R, and Z_T to calculate and record the output voltage magnitude in decibels from the equation V_{out} (dB) $= 20 \log [V_S (R / Z_T)]$.

    d. Use R and X_L to calculate and record the output voltage phase angle $\varphi_{out} = -\tan^{-1} (X_L / R)$.

Table 12-3: Calculated Values for Low-Pass Filter Circuit ($R = 5.1$ kΩ, $L = 16$ mH)

f	X_L	Z_T	V_{out} (dB)	φ_{out}	f	X_L	Z_T	V_{out} (dB)	φ_{out}
1 kHz					100 kHz				
2 kHz					200 kHz				
5 kHz					500 kHz				
10 kHz					1 MHz				
20 kHz					2 MHz				
50 kHz					5 MHz				

3) Double-click the Bode plotter to expand it.

4) Set the Bode plotter settings to those in Table 12-4.

Table 12-4: Low-Pass Filter Settings for the Bode Plotter

Mode	Horizontal	Vertical	Mode	Horizontal	Vertical
Magnitude	Log F: 10 MHz I: 500 Hz	Log F: 10 dB I: −50 dB	Phase	Log F: 10 MHz I: 500 Hz	Lin F: 0 deg I: −90 deg

5) Run the simulation.

6) When the Bode plotter has finished collecting data, stop the simulation.

7) For each of the frequencies in Table 12-5:

    a. Use the cursor in the Bode plotter to measure the magnitude of V_R. Record this value as V_R.

    b. Use the cursor in the Bode plotter to measure the phase of V_R. Record this value as φ_R.

Note that you can enter frequencies using "k" and "M" to specify kilohertz and megahertz, respectively (e.g., "10k" is the same as "10000").

Table 12-5: Measured Data for Low-Pass *RL* Filter

f	V_R	φ_R	f	V_R	φ_R	f	V_R	φ_R
1 kHz			20 kHz			500 kHz		
2 kHz			50 kHz			1 MHz		
5 kHz			100 kHz			2 MHz		
10 kHz			200 kHz			5 MHz		

Observations:

12.8 High-Pass *RL* Filter

1) Open the Multisim file *DC_AC_Exp_12_Part_02b*.

2) For each of the frequencies in Table 12-6:

    a. Use f and L to calculate and record the inductive reactance $X_L = 2\pi f L$.

    b. Use R and X_L to calculate and record the impedance magnitude $Z_T = \sqrt{R^2 + X_L^2}$.

    c. Use V_S, R, and Z_T to calculate and record the output voltage magnitude in decibels from the equation V_{out} (dB) $= 20 \log [V_S (X_L / Z_T)]$.

    d. Use R and X_L to calculate and record the output voltage phase angle $\varphi_{out} = 90° - \tan^{-1} (X_L / R)$.

Table 12-6: Calculated Values for High-Pass *RL* Filter Circuit (R = 13 kΩ, L = 200 mH)

f	X_L	Z_T	V_{out}(dB)	φ_{out}	f	X_L	Z_T	V_{out}(dB)	φ_{out}
100 Hz					10 kHz				
200 Hz					20 kHz				
500 Hz					50 kHz				
1 kHz					100 kHz				
2 kHz					200 kHz				
5 kHz					500 kHz				

3) Double-click the Bode plotter to expand it.

4) Set the Bode plotter settings to those in Table 12-7.

Table 12-7: High-Pass Filter Settings for the Bode Plotter

Mode	Horizontal	Vertical	Mode	Horizontal	Vertical
Magnitude	Log F: 1 MHz I: 50 Hz	Log F: 10 dB I: −50 dB	Phase	Log F: 1 MHz I: 50 Hz	Lin F: 90 deg I: 0 deg

5) Run the simulation.
6) When the Bode plotter has finished collecting data, stop the simulation.
7) For each of the frequencies in Table 12-8:
   a. Use the Bode plotter to measure the magnitude of V_R. Record this value as V_R.
   b. Use the Bode plotter to measure the phase of V_R. Record this value as φ_R.

Table 12-8: Measured Data for High-Pass *RL* Filter

f	V_R	φ_R	f	V_R	φ_R	f	V_R	φ_R
100 Hz			2 kHz			50 kHz		
200 Hz			5 kHz			100 kHz		
500 Hz			10 kHz			200 kHz		
1 kHz			20 kHz			500 kHz		

Observations:

Conclusions for Part 2

Questions for Part 2

1) The corner frequency f_c is the frequency at which the reactance in a circuit is equal to the resistance. What are the corner frequencies for the *RL* high-pass and low-pass filters?

2) What is the approximate magnitude in decibels and phase in degrees for the high-pass and low-pass filters at their corner frequencies?

3) How does the experimental Bode plot data for the low-pass filter compare with the idealized Bode plot of Figure 12-1 at $1/10 f_c$, f_c, and $10 f_c$?

Name _____   Class _____

Date _____   Instructor_____

DC/AC Experiment 13 - Series and Parallel Resonance

13.1 Introduction

Whenever capacitance or inductance is present in a circuit, there will be a reactive component that introduces a phase shift between the applied voltage and the current in the circuit. When the circuit contains both capacitance and inductance, the capacitive reactance and inductive reactance act in opposition to each other. When a capacitor and inductor are in series, the capacitor voltage lags the current by 90° and the inductor voltage leads the current by 90° so that the voltages tend to cancel, thereby reducing the total voltage across both. When a capacitor and inductor are in parallel, the capacitor current leads the applied voltage by 90° and the inductor current lags the applied voltage by 90° so that the currents tend to cancel, thereby reducing the total current. When the capacitive reactance and inductive reactance are equal, a phenomenon called **resonance** occurs in which the two opposing reactive effects cancel completely. In an ideal series circuit the capacitive reactance and inductive reactance are exactly equal and opposite, so that the only opposition to current flow is the resistance in the circuit. In an ideal parallel circuit the capacitor current and inductor current are equal and opposite and cancel. Therefore, from Kirchhoff's current law, no current can enter or leave the parallel combination (called the tank) and the opposition to current flow is infinite. The frequency at which the reactances are equal is called the **resonance frequency**, and the circuit is said to be at resonance. In addition to the resonant frequency, two related characteristics associated with resonance are the circuit bandwidth and quality factor, or Q. The bandwidth and Q indicate how rapidly the circuit characteristics change as the frequency approaches the resonant frequency.

When a circuit is at resonance it exhibits some interesting properties. Although the inductor and capacitor voltages in a series circuit at resonance cancel each other so that the net voltage is zero, the individual inductor voltage and capacitor voltage can be quite large, even larger than the applied source voltage! Similarly, although the inductor and capacitor currents in an ideal parallel circuit at resonance cancel, the current through the capacitor and inductor (called the tank current) can also be extremely large. In both cases, the circuit appears purely resistive.

In Part 1 of this experiment, you will use the oscilloscope to calculate, observe, and measure how the current magnitude and phase change in a series *RLC* circuit (a circuit containing resistance, inductance, and capacitance) as the frequency approaches and passes through the resonant frequency. In Part 2, you will measure the resonant frequency and bandwidth of series and parallel *RLC* circuits, determine the effect of circuit resistance on the circuit Q, and observe the inductor and capacitor voltages and currents at resonance.

13.2 Reading

Floyd and Buchla, *Electric Circuits Fundamentals, 8th Edition*, Chapter 13.

13.3 Key Objectives

Part 1:   Use the oscilloscope to measure and record the current magnitude and phase in a series *RLC* circuit as frequency changes.

Part 2:   Use the Bode plotter to determine the resonant frequency, bandwidth, and Q for series and parallel *RLC* circuits. Calculate and verify the capacitor and inductor voltages in a series and parallel *RLC* circuit at resonance.

13.4 Multisim Files

Part 1:   *DC_AC_Exp_13_Part_01*

Part 2:   *DC_AC_Exp_13_Part_02a* through *DC_AC_Exp_13_Part_02d*.

Part 1:   Magnitude and Phase of Series *RLC* Circuits

1)   Open the Multisim file *DC_AC_Exp_13_Part_01*.

2)   For each of the frequencies in Table 13-1:

a. Use f, L and C to calculate and record $X_L = 2\pi f L$ and $X_C = 1 / (2\pi f C)$.

b. Use X_L and X_C to calculate and record the total reactance $X_T = X_L - X_C$.

c. Use R and X_T to calculate and record the impedance magnitude $Z_T = \sqrt{R^2 + X_T^{\,2}}$.

d. Use V_S, R, and Z_T to calculate and record the resistor voltage magnitude $V_R = V_S (R / Z_T)$.

e. Use R and X_T to calculate and record the resistor voltage phase angle $\varphi_R = -\tan^{-1}(X_T / R)$.

Table 13-1: Calculated Series _RLC_ Values ($R = 10\ \Omega$, $L = 5.1$ mH, $C = 310$ nF)

f (kHz)	X_L	X_C	X_T	Z_T	V_R	φ_R
1.0						
2.0						
3.0						
4.0						
5.0						
6.0						
7.0						
8.0						
9.0						
10.0						

3) For each of the rows in Table 13-2:

a. Set the voltage source V_S to the indicated frequency.

b. Calculate and record to three significant digits the period $T = 1 / f$.

c. Measure and record to three significant digits the peak amplitude of the Channel B signal V_R.

d. Start the simulation and allow the voltages on each channel to settle so that the positive and negative peaks are equal.

e. Measure to three significant digits the time difference $\Delta t = t_{\text{CH A}} - t_{\text{CH B}}$ between corresponding points (positive peak, negative peak, zero crossing, etc.) of Channel A and Channel B.

f. Calculate and record to three significant digits the phase shift $\varphi_R = (\Delta t / T) \times 360°$ between Channel A and Channel B.

Table 13-2: Measured Series _RLC_ Values

f (kHz)	T	V_R	Δt	φ_R	f (kHz)	T	V_R	Δt	φ_R
1.0					6.0				
2.0					7.0				
3.0					8.0				
4.0					9.0				
5.0					10.0				

Observations:

Conclusions for Part 1

Questions for Part 1

1) How do you explain the positive phase angles for V_R for lower frequencies and the negative phase angles at higher frequencies?

2) From the calculations in Table 13-1, at what frequency are the capacitive and inductive reactances equal? What does this signify for the circuit at that frequency?

3) From the data in Table 13-2, what characteristics do total current and resistor voltage exhibit at resonance?

Part 2:   Series and Parallel Resonance

13.5 Bandwidth and Q of Series Resonant Circuits

1) Open the Multisim file *DC_AC_Exp_13_Part_02a*.
2) Double-click on the Bode plotter to expand it.
3) Calculate the resonant frequency f_r from the equation

$$f_r = \frac{1}{2\pi\sqrt{LC}}$$

$f_r =$

4) For each of the rows in Table 13-3:
   a. Change the resistor value to the specified value of R.
   b. Run the simulation. When the Bode plotter finishes collecting data, stop the simulation.
   c. Measure and record the frequency f_r for the maximum magnitude.
   d. Measure and record the lower critical frequency f_{CL} for which the magnitude is 3 dB less than the maximum magnitude.
   e. Measure and record the upper critical frequency f_{CH} for which the magnitude is 3 dB less than the maximum magnitude.
   f. Use the values of f_{CL} and f_{CH} to calculate the bandwidth $BW = f_{CH} - f_{CL}$.
   g. Use the values of f_r and BW to calculate the circuit $Q = f_r / BW$.

Table 13-3: Series Resonance, Bandwidth, and *Q*

R	*f_r*	*f_{CL}*	*f_{CH}*	*BW*	*Q*
10 Ω					
20 Ω					
51 Ω					
100 Ω					

How do the bandwidth and *Q* change as the series resistance of the circuit increases?

13.6 Voltages in Series Resonant *RLC* Circuits

1) Open the Multisim file *DC_AC_Exp_13_Part_02b*.
2) Calculate the resonant frequency for the circuit.

   $f_r =$

3) Set the frequency of V_S to your calculated resonant frequency.
4) Set XMM_1 and XMM_2 to ac voltmeter mode and XMM_3 to ac ammeter mode.
5) Calculate the reactive and inductive reactance at the resonant frequency.

   X_C:

   X_L:

6) For each of the rows in Table 13-4:
   a. Use V_S and R to calculate and record the total current $I_T = V_S / R$ at resonance.
   b. Use I_T and X_C to calculate and record the capacitor voltage $V_C = I_T X_C$ at resonance.
   c. Use I_T and X_L to calculate and record the inductor voltage $V_L = I_T X_L$ at resonance.
7) For each of the rows in Table 13-4:
   a. Set the resistor value to the specified value of R.
   b. Start the simulation.
   c. Measure and record to three significant digits the current $I_{T\,(MEAS)}$ through R, the voltage $V_{C\,(MEAS)}$ across C, and the $V_{L\,(MEAS)}$ across L.

Table 13-4: Series *RLC* Circuit Values at Resonance

R	I_T	V_C	V_L	$I_{T\,(MEAS)}$	$V_{C\,(MEAS)}$	$V_{L\,(MEAS)}$
1.0 kΩ						
510 Ω						
100 Ω						
51 Ω						
10 Ω						

Observations:

13.7 Bandwidth and *Q* of Parallel Resonant Circuits

1) Open the Multisim file *DC_AC_Exp_13_Part_02c*.
2) Double-click on the Bode plotter to expand it.
3) Calculate the resonant frequency f_r from the equation

$$f_r = \frac{1}{2\pi\sqrt{LC}}$$

$f_r =$

4) For each of the rows in Table 13-5:
   a. Change the resistor value to the specified value of *R*.
   b. Run the simulation. When the Bode plotter finishes collecting data, stop the simulation.
   c. Measure and record the frequency f_r for the maximum magnitude.
   d. Determine and record the lower critical frequency f_{CL} for which the magnitude is 3 dB less than that for the maximum magnitude.
   e. Determine and record the lower critical frequency f_{CH} for which the magnitude is 3 dB less than that for the maximum magnitude.
   f. Use the values of f_{CL} and f_{CH} to calculate and record the bandwidth $BW = f_{CL} - f_{CH}$.
   g. Use the values of f_r and BW to calculate and record the circuit $Q = f_r / BW$.

Table 13-5: Parallel Resonance, Bandwidth, and *Q*

R	f_r	f_{CL}	f_{CH}	*BW*	*Q*
100 Ω					
200 Ω					
510 Ω					
1.0 kΩ					

How do the bandwidth and *Q* change as the series resistance of the circuit increases?

13.8 Currents in Parallel Resonant *RLC* Circuits

1) Open the Multisim file *DC_AC_Exp_13_Part_02d*.
2) Calculate the resonant frequency for the circuit.

$f_r =$

3) Set the frequency of V_S to your calculated resonant frequency.
4) Calculate the capacitive and inductive reactance at the resonant frequency.

X_C:

X_L:

5) Calculate the current through the capacitor and inductor at resonance.

I_C: $V_1 / X_C =$

I_L: $V_1 / X_L =$

6) Double-click the multimeters to expand them and set them to ac ammeter mode.

7) Start the simulation and record to three significant figures the current values for I_T, I_L, and I_C.

I_T:

I_C:

I_L:

Observations:

Conclusions for Part 2

Questions for Part 2

1) What is the danger of operating an *RLC* circuit at its resonant frequency?

2) If you wanted to increase the bandwidth of a series *RLC* circuit without affecting the resonant frequency, how you could do so?

3) One interpretation of the *Q* of a circuit is the ability of a circuit to retain stored energy. How would this explain the effect of resistor values you observed for series and parallel *RLC* circuits?

DC/AC Experiment 14 - Transformers

14.1 Introduction

Previous experiments introduced two devices whose operations are based on electromagnetism: the relay and the inductor. The current in a relay coil creates a magnetic field that opens or closes switch contacts on a movable armature, whereas the current through an inductor creates and stores energy in a magnetic field. This experiment examines a third electromagnetic device, called a *transformer*. Physically a basic transformer consists of two conductive coils wound on a common core. AC current flowing through one coil, called the primary, creates a changing magnetic field. These lines of magnetic force pass through and induce a voltage across the second coil, called the secondary. In an ideal transformer all the lines of magnetic force from one coil pass through the other coil so that the coefficient of coupling $k = 1$. For many practical transformers k is very close to 1. The Multisim circuits for this experiment will use an ideal transformer model for which $k = 1$.

Transformers typically use a turns ratio $n = N_{sec} / N_{pri}$, that indicates the ratio of the number of windings on the secondary to the number of windings on the primary. The Multisim ideal transformer for this experiment uses the primary and secondary inductances. You can relate the desired transformer primary and secondary voltages from the equation that relates the inductance L to the number of turns N, core permeability μ, core cross-sectional area A, and flux path l:

$$\frac{N^2 \mu A}{l} = L \text{ (from the equation for inductance)}$$

$$N^2 = L\left(\frac{l}{\mu A}\right)$$

$$N = \sqrt{L\left(\frac{l}{\mu A}\right)}$$

For a transformer, $V_{pri} / V_{sec} = N_{pri} / N_{sec}$.

$$\frac{V_{pri}}{V_{sec}} = \frac{\sqrt{L_{pri}\left(\dfrac{l_{pri}}{\mu_{pri} A_{pri}}\right)}}{\sqrt{L_{sec}\left(\dfrac{l_{sec}}{\mu_{sec} A_{sec}}\right)}}$$

When both coils are wound on the same uniform core, $l / \mu A$ is the same for both coils. Therefore

$$\frac{V_{pri}}{V_{sec}} = \sqrt{\frac{L_{pri}}{L_{sec}}} \Rightarrow V_{pri} = V_{sec}\sqrt{\frac{L_{pri}}{L_{sec}}} \Rightarrow V_{sec} = V_{pri}\sqrt{\frac{L_{sec}}{L_{pri}}}$$

$$\left(\frac{V_{pri}}{V_{sec}}\right)^2 = \frac{L_{pri}}{L_{sec}} \Rightarrow L_{pri} = L_{sec}\left(\frac{V_{pri}}{V_{sec}}\right)^2 \Rightarrow L_{sec} = L_{pri}\left(\frac{V_{sec}}{V_{pri}}\right)^2$$

To set up a desired transformer step-up or step down ratio, simply select the appropriate equation, substitute in the required values, and solve for the necessary missing value. Note that, although a transformer can step up the secondary voltage, it does not amplify power. If a transformer steps up the secondary voltage, it steps down the secondary current, and conversely, so that the primary and secondary power are (ideally) equal.

Transformers have three major applications. The first application is to change the amplitude of an ac voltage or current to a higher or lower value. The second application is to match the impedance of a load to the equivalent source impedance for maximum power transfer from the source to the load. The third application is to provide electrical isolation, as transformers do not couple, or pass, dc values from the primary to secondary coil and allow circuits on the secondary to be isolated from the ground reference on the primary.

In Part 1 of this experiment, you will investigate the operation of step-up transformers (transformers in which the secondary voltage is larger than the primary voltage) and step-down transformers (transformers in which the secondary voltage is smaller than the primary voltage). In Part 2, you will examine the effects of using transformers to match source and load impedances.

14.2 Reading

Floyd and Buchla, *Electric Circuits Fundamentals, 8th Edition*, Chapter 14

14.3 Key Objectives

Part 1: Calculate and verify the transformer primary and secondary characteristics required for specific step-up and step-down transformer ratios.

Part 2: Discuss the advantage of adding transformer impedance matching to circuits.

14.4 Multisim Files

Part 1: *DC_AC_Exp_14_Part_01*

Part 2: *DC_AC_Exp_14_Part_02a* and *DC_AC_Exp_14_Part_02b*

Part 1: Voltage Conversion

14.5 Step-Down Transformer Circuits

1) For each of the rows in Table 14-1, use the specified values to calculate and record to three significant digits the missing value of V_{pri}, V_{sec}, L_{pri}, or L_{sec}.

2) Open the Multisim file *DC_AC_Exp_14_Part_01*.

3) Double-click on the DMMs and set them for ac voltmeter mode.

4) For each of the rows in Table 14-1:

    a. Set the amplitude of V_S to the specified value of V_{pri}.

    b. Double-click on T_1 to open the transformer **Properties** window.

    c. Set the **Primary Coil Inductance** to the specified value of L_{pri}.

    d. Set the **Secondary Coil Inductance** to the specified value of L_{sec}.

    e. Click the **OK** button to accept the changes.

    f. Start the simulation.

    g. Record the measured XMM_1 value as $V_{pri\,(MEAS)}$.

    h. Record the measured XMM_2 value as $V_{sec\,(MEAS)}$.

    i. Stop the simulation.

Table 14-1: Step-Down Transformer Values

V_{pri}	V_{sec}	L_{pri}	L_{sec}	$V_{pri\,(MEAS)}$	$V_{sec\,(MEAS)}$
	15 V$_{rms}$	250 mH	10 mH		
15 V$_{rms}$		100 mH	20 mH		
100 V$_{rms}$	20 V$_{rms}$		25 mH		
120 V$_{rms}$	12 V$_{rms}$	1.0 H			

Observations:

14.6 Step-Up Transformer Circuits

1) For each of the rows in Table 14-2, use the specified values to calculate and record to three significant digits the missing value of V_{pri}, V_{sec}, L_{pri}, or L_{sec}.

2) Open the Multisim file *DC_AC_Exp_14_Part_01*.

3) Double-click on the DMMs and set them for ac voltmeter mode.

4) For each of the rows in Table 14-2:

    a. Set the amplitude of V_S to the specified value of V_{pri}.

    b. Double-click on T_1 to open the transformer **Properties** window.

    c. Set the **Primary Coil Inductance** to the specified value of L_{pri}.

    d. Set the **Secondary Coil Inductance** to the specified value of L_{sec}.

    e. Click the **OK** button to accept the changes.

    f. Start the simulation.

    g. Record the measured XMM_1 value as $V_{pri\,(MEAS)}$.

    h. Record the measured XMM_2 value as $V_{sec\,(MEAS)}$.

    i. Stop the simulation.

Table 14-2: Step-Up Transformer Values

V_{pri}	V_{sec}	L_{pri}	L_{sec}	$V_{pri\,(MEAS)}$	$V_{sec\,(MEAS)}$
	1.8 V$_{rms}$	15 mH	3.375 H		
24 V$_{rms}$		12 mH	1.2 H		
6 V$_{rms}$	72 V$_{rms}$		7.2 H		
12 V$_{rms}$	48 V$_{rms}$	10 mH			

Observations:

Conclusions for Part 1

Questions for Part 1

1) A transformer with a uniform core (i.e., the permeability μ and cross-sectional area A is the same throughout the core) has $L_{pri} = 250$ mH and $L_{sec} = 500$ mH. Is the transformer a step-up or step-down transformer?

2) You wish to have $V_{sec} = 5.0$ V when $V_{pri} = 12.0$ V for a transformer with $L_{sec} = 10$ mH. What must L_{pri} be?

Part 2: Impedance Matching

1) Open the Multisim file *DC_AC_Exp_14_Part_02a*.
2) Double-click the wattmeters to expand them.
3) For each row in Table 14-3:
   a. Calculate and record to three significant digits the total current $I_T = V_S / (R_S + R_L)$.
   b. Calculate and record to three significant digits the input power $P_{in} = V_S I_T$.
   c. Calculate and record to three significant digits the power $P_{RL} = I_T^2 R_L$ dissipated by R_L.
4) For each row in Table 14-3:
   a. Change the value of R_L to the specified value of R_L.
   b. Start the simulation.
   c. Record to three significant digits the value of XWM_1 as $P_{in\,(MEAS)}$.
   d. Record to three significant digits the value of XWM_2 as $P_{RL\,(MEAS)}$.
   e. Stop the simulation.

Table 14-3: Unmatched Impedance Circuit Values

R_L	I_T	P_{in}	P_{RL}	$P_{in\,(MEAS)}$	$P_{RL\,(MEAS)}$
1 Ω					
2 Ω					
5 Ω					
10 Ω					
20 Ω					
50 Ω					
75 Ω					
100 Ω					
200 Ω					

Observations:

5) For each row in Table 14-4:
   a. Change the value of R_L to the specified value of R_L.
   b. Using the specified value of L_{pri}, calculate and record to four significant digits the necessary value of L_{sec} to impedance match R_L to R_S from the equation
   $$L_{sec} = L_{pri} (R_L / R_S)$$
   c. Change the value of L_{sec} to your calculated value.
   d. Start the simulation.

e.   Record to three significant digits the value of XWM_1 as $P_{in\,(MEAS)}$.

f.   Record to three significant digits the value of XWM_2 as $P_{RL\,(MEAS)}$.

g.   Stop the simulation.

Table 14-4: Matched Impedance Circuit Values

R_L	L_{pri}	L_{sec}	$P_{in\,(MEAS)}$	$P_{RL\,(MEAS)}$
1 Ω	1 H			
2 Ω	1 H			
5 Ω	1 H			
10 Ω	1 H			
20 Ω	1 H			
50 Ω	1 H			
75 Ω	1 H			
100 Ω	1 H			
200 Ω	1 H			

Observations:

Conclusions for Part 2

Questions for Part 2

1)   How can you prove that $L_{sec} = L_{pri}\,(R_L\,/\,R_S)$ for matching impedance with the ideal transformer?

2) If a transformer has L_{pri} = 50 mH, what L_{sec} value is necessary to match an 8 Ω load to a 75 Ω source?

DC/AC Experiment 15 - Time Response of Reactive Circuits

15.1 Introduction

In previous experiments you examined the behavior of reactive circuits, including the effect of time constants on the transient (short-term) behavior of RC and RL circuits and the steady-state amplitude and frequency response to ac (sinusoidal) signals. This experiment further investigates the time response of reactive circuits by examining their response to single or multiple pulses. This behavior depends upon the width of the pulses compared to the time constant τ of the circuit. An RC or RL circuit requires 5τ to fully respond to an input change so that the reactive component can fully energize and de-energize. In general, the responses can be classified as three general categories, depending upon the relation of the pulse width t_w to 5τ.

If $5\tau > t_w$, the output of a low-pass circuit resembles that of an integrator and the output of a high-pass circuit resembles the input. The output of an integrator is proportional to the total area under the input waveform. If the input voltage is a dc voltage, the output of an integrator will be a linear ramp, as the total area under the dc voltage increases linearly with time.

If $5\tau = t_w$, the reactive component can fully charge (or energize) and discharge (or de-energize) so that the output reaches its maximum and minimum values.

If $5\tau < t_w$, the output of a low-pass circuit resembles the input and the output of a high-pass circuit resembles that of a differentiator. The output of a differentiator is proportional to how rapidly the input is changing. If the input voltage is a constant voltage, the output of a differentiator is zero as the voltage is not changing.

In Part 1 of this experiment, you will observe and describe how the output of integrators and differentiators change in response to a single pulse as the pulse width changes relative to the circuit time constant. In Part 2, you will examine and describe how the output of low-pass and high-pass circuits change in response to multiple pulses.

15.2 Reading

Floyd and Buchla, *Electric Circuits Fundamentals, 8th Edition*, Chapter 15

15.3 Key Objectives

Part 1: Observe and describe the output of integrators and differentiators as the width of a single pulse changes relative to the circuit time constant.

Part 2: Observe and describe the output of low-pass and high-pass circuits for multiple pulses for which $5\tau > t_w$.

15.4 Multisim Files

Part 1: *DC_AC_Exp_15_Part_01a* and *DC_AC_Exp_15_Part_01b*

Part 2: *DC_AC_Exp_15_Part_02a* and *DC_AC_Exp_15_Part_02b*

Part 1: Response of Reactive Circuits to Single Pulses

15.5 The *RL* Integrator

1) For each of the rows in Table 15-1,

a. Use the values of t_w and N to calculate and record to three significant digits the value of $5\tau = N \times t_w$. N, which is the ratio $5\tau / t_w$, relates the time required for an inductor or capacitor to fully charge (or energize) or discharge (or de-energize) to the width of the applied pulse.

b. Use the value of 5τ to calculate and record to three significant digits the value of $\tau = 5\tau / 5$.

c. Use the values of L and τ to calculate and record to three significant digits the value of $R = L / \tau$.

2) Open the Multisim file *DC_AC_Exp_15_Part_01a*.

3) Double-click the oscilloscope to expand it.

4) For each of the rows in Table 15-1:

    a. Change the value of *R* to the calculated value.

    b. Start the simulation, allow the waveform to settle, and stop the simulation.

    c. Sketch the Channel B waveform in Plot 15-1 in the graph corresponding to the value of *N*.

Table 15-1: Calculated Values for *RL* Integrator

t_w	N	5τ	τ	L	R
400 μs	0.1			100 mH	
400 μs	1			100 mH	
400 μs	10			100 mH	

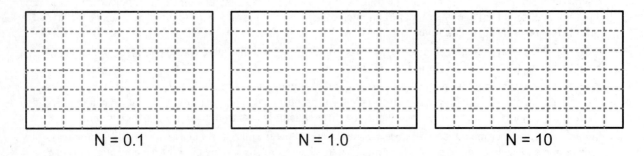

N = 0.1   N = 1.0   N = 10

Plot 15-1: *RL* Integrator Waveforms

15.6 The *RC* Differentiator

1) For each of the rows in Table 15-2:

    a. Use the values of t_w and *N* to calculate and record to three significant digits the value of $5\tau = N \times t_w$.

    b. Use the value of 5τ to calculate to three significant digits the value of $\tau = 5\tau / 5$.

    c. Use the values of *C* and τ to calculate to three significant digits the value of $R = \tau / C$.

2) Open the Multisim file *DC_AC_Exp_15_Part_01b*.

3) Double-click the oscilloscope to expand it.

4) For each of the rows in Table 15-2:

    a. Change the value of *R* to the calculated value.

    b. Start the simulation, allow the waveform to settle, and stop the simulation.

    c. Sketch the Channel B waveform in Plot 15-2 in the graph corresponding to the value of *N*.

Table 15-2: Calculated Values for *RC* Differentiator

t_w	N	5τ	τ	C	R
400 μs	0.1			50 nF	**160 Ω**
400 μs	1			50 nF	**1.60 kΩ**
400 μs	10			50 nF	**16.0 kΩ**

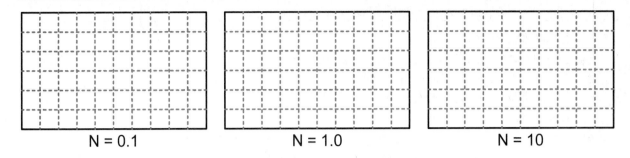

Plot 15-2: *RC* **Differentiator Waveforms**

Conclusions for Part 1

Questions for Part 1

1) For the integrator, will the output most closely resemble the input for a longer or shorter time constant?

2) For the circuits in this part of the experiment, what are the average values of the integrator and differentiator after the waveforms settle?

3) How do you explain the positive and negative spikes for $N = 0.1$ for the differentiator output?

Part 2: Response of Reactive Circuits to Multiple Pulses

15.7 The *RC* Integrator

1) Open the Multisim file *DC_AC_Exp_15_Part_02a*.
2) Calculate the time constant for the circuit.

$$\tau = RC =$$

3) Run the simulation.
4) After the oscilloscope collects one screen of data, stop the simulation.
5) Compare the output waveform with that of the *RL* integrator.

Observations:

15.8 The *RL* Differentiator.

1) Open the Multisim file *DC_AC_Exp_15_Part_02b*.

2) Calculate the time constant for the circuit.

   $\tau = L / R =$ **39 mH / 390 Ω = 100 μs.**

3) Run the simulation.

4) After the oscilloscope collects one screen of data, stop the simulation.

5) Compare the output waveform with that of the *RC* differentiator.

   Observations:

Conclusions for Part 2

Questions for Part 2

1) For the circuits in this part of the experiment, why does it take several pulses for output waveforms of the integrator and differentiator to settle?

2) For the *RC* integrator in *DC_AC_Exp_15_Part_02a*, approximately how many pulses does it take for the output to settle to its steady-state value? How does this compare with the time constant τ of the circuit?

3) For the *RL* differentiator in *DC_AC_Exp_15_Part_02b*, approximately how many pulses does it take for the output to settle to its steady-state value? How does this compare with the time constant τ of the circuit?

Digital Experiments

Digital Experiment 1 - Introduction to Multisim for Digital Circuits............113

Digital Experiment 2 - Number Systems 123

Digital Experiment 3 - Logic Gates..129

Digital Experiment 4 - Boolean Algebra and DeMorgan's Theorem 137

Digital Experiment 5 - Combinational Logic 141

Digital Experiment 6 - The Full Adder...147

Digital Experiment 7 - Encoders, Decoders, and Parity Checkers 153

Digital Experiment 8 - Latches and Flip-Flops.....................................161

Digital Experiment 9 - Counters 171

Digital Experiment 10 - Shift Registers 177

Digital Experiment 11 - Memory Devices and Operation 181

Digital Experiment 12 - Programmable Logic Concepts..........................185

Digital Experiment 13 - Signal Interfacing 191

Digital Experiment 14 - Processor Support Circuitry 197

Digital Experiment 15 - The Arithmetic Logic Unit...............................203

Digital Experiment 1 - Introduction to Multisim for Digital Circuits

1.1   Introduction

Electronic circuits consist of two general classes. The first class, called *analog circuits*, are circuits in which electrical quantities can have any value over a range. The second class, called *digital circuits*, are circuits in which electrical quantities can have only specific values. To illustrate the concepts of analog and digital, compare using a dimmer switch and toggle switch to control the lights in a room. A dimmer switch represents analog control, as it allows you to adjust the brightness of the lights to whatever level you wish. A toggle switch represents digital control, as it allows you to set the lights to only two specific states: *on* or *off*. The most common digital circuits are *binary* digital circuits, which means that each signal has only two valid states.

Although you can use multimeters and oscilloscopes to observe the operation of digital circuits, you will typically use these instruments only when troubleshooting. This is because signal logic levels (HIGH or LOW) and relative timing between signals, rather than exact amplitude and timing information, are important for normal operation of digital circuits. Special instruments, such as logic probes and logic analyzers, exist for measuring logic levels and signal timing in digital circuits. Logic analyzers use a special digital signal, called a *clock*, to coordinate the time at which it takes samples. You will learn more about clock signals in this experiment.

In Part 1 of this experiment, you will familiarize yourself with the Multisim tools for observing digital circuits. In Part 2, you will use the logic probe to examine the operation of some digital circuits. In Part 3, you will use the logic analyzer to observe multiple logic signals simultaneously.

1.2   Reading

National Instruments, *NI Circuit Design Suite: Getting Started with Circuit Design Suite*, Chapters 1 and 2

1.3   Key Objectives

Part 1:   Describe the common Multisim instruments and components associated with digital circuits.

Part 2:   Show how to access and use the Multisim logic probes.

Part 3:   Show how to access and use the Multisim logic analyzer to observe the operation of a digital circuit.

1.4   Multisim Files

Part 2:   *Digital_Exp_01*

Part 3:   *Digital_Exp_01*

Part 1:   Multisim Digital Tools

In this part of the experiment you will learn the tools that you will use to work with most digital circuits. For details on using Multisim or the Multisim interface, refer to DC/AC Experiment 1 in this manual.

1.5   Multisim Toolbars

1.5.1   The Component Toolbar

The **Component Toolbar**, shown in Figure 1-1, contains tools that let you access various components with which to create and analyze digital circuits.

Figure 1-1: The Component Toolbar

The component tools with which you will most often work for basic digital electronics are the **Place Source**, **Place TTL**, **Place CMOS**, and **Place Indicator** tools, shown in Figure 1-2.

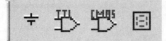

Figure 1-2: Place Source, Place TTL, Place CMOS, and Place Indicator Tools

1.5.2 The Instruments Toolbar

The **Instruments Toolbar**, shown in Figure 1-3, provides instruments with which you can measure and evaluate the operation of circuits. Some instruments, like the multimeter, oscilloscope, and logic analyzer, are real devices that technicians and engineers use to analyze real-world circuits. Other instruments, like the Bode plotter and logic converter, exist only within the Multisim application and are convenient tools for you to simulate, analyze, and debug your circuit designs.

Figure 1-3: The Instruments Toolbar

The instrument tools with which you will most often work for digital electronics are the **Function Generator**, **Frequency Counter**, **Word Generator**, **Logic Analyzer**, and **Logic Converter** tools, shown in Figure 1-4.

Figure 1-4: Common Digital Tools

1.6 Identifying Multisim Digital Tools

1) Start the Multisim program.

2) Use your cursor and the Multisim tooltip to identify each of the digital tools shown in Table 1-1.

Table 1-1: Multisim Digital Tools (continued on next page)

Tool Graphic	Associated Toolbar	Tool Identity

Tool Graphic	Associated Toolbar	Tool Identity

Questions for Part 1

1) Would the mercury thermometer with the numbered scale shown in Figure 1-5 be considered analog or digital in nature? Why?

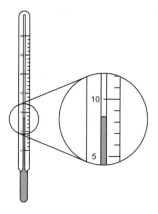

Figure 1-5: Mercury Thermometer

2) Would you probably design a digital circuit or analog circuit to grade multiple-choice tests? Why?

Part 2:   Digital Measurements in Multisim

In this part of the experiment you will learn to use logic probes to observe the operation of a typical digital circuit.

1.7   Using Probes

In Multisim, probes are visual indicators that allow you to determine whether the state of a signal line is HIGH or LOW. When the probe is lit, the logic state is HIGH, and when the probe is not lit, the logic state is LOW.

1) Open the file *Digital_Exp_01*.

2) Select the **Place Indicator** tool on the **Component Toolbar**. This will open the component browser with the indicator components, shown in Figure 1-6.

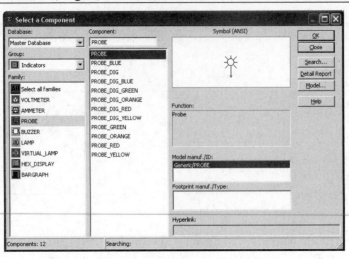

Figure 1-6: Indicator Component Browser

3) Select "Probe" from the **Family:** window.

4) Select "PROBE_DIG_RED" from the **Component:** list.

5) Click the **OK** button and place the probe above Q_0 junction, as shown in Figure 1-7.

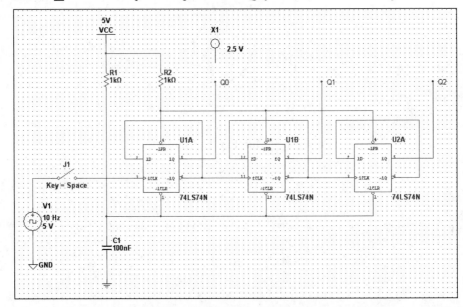

Figure 1-7: Placing the First Probe

As you can see, the probe is a very simple device. When you connect a probe into a circuit, it measures and compares the voltage on its pin to a threshold value (in this case, the default value of 2.5 V). If the signal voltage is greater than the threshold value the probe lights up, indicating a HIGH. If the signal voltage is less than the threshold value, the probe does not light up, indicating a LOW.

6) Connect the probe X_1 to Q_0. Note that the junction for Q_0 disappears when you do so.

7) Connect a "PROBE_DIG_GREEN" to Q_1.

8) Connect a "PROBE_DIG_BLUE" to Q_2. The circuit should now look like that in Figure 1-8.

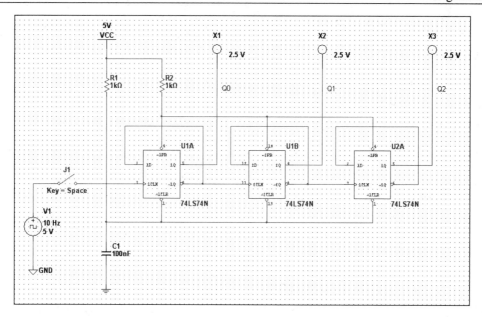

Figure 1-8: Circuit with Probes Installed

9) Start the simulation.

10) Press the space bar to close switch J_1. This connects the 10 Hz clock signal to the circuit.

11) When the state of the probes changes, press the space bar again to open switch J_1.

12) Record the state of the output for each probe in Table 1-2. If the probe is not lit, write "LOW". If the probe is lit, write "HIGH".

13) Repeat Steps 9 through 11 until you have completed the table.

14) Stop the simulation and close the circuit without saving.

Table 1-2: Probe Circuit States

State	Q_0 (RED)	Q_1 (GREEN)	Q_2 (BLUE)
1			
2			
3			
4			
5			
6			
7			
8			
9			
10			

Conclusions for Part 2

Questions for Part 2

1) From Table 1-2, how many unique states does the circuit appear to have?

2) From Table 1-2, does there appear to be any pattern to how the state of each probe changes?

3) If the frequency of the clock had been 10 kHz rather than 10 Hz, so that the probes changed state one thousand times faster, how easily could you have completed Table 1-2?

Part 3:   Using the Logic Analyzer

As you should have determined Part 2 of the experiment, probes are a convenient way to determine the logic state of signal lines but have limited value at high frequencies. The logic analyzer also allows you to determine the logic state of signal lines but can operate at much higher frequencies.

1) Open the file *Digital_Exp_01*.

2) Select the **Logic Analyzer** tool on the **Instruments Toolbar**.

3) Place the logic analyzer above and to the right of the circuit, as shown in Figure 1-9.

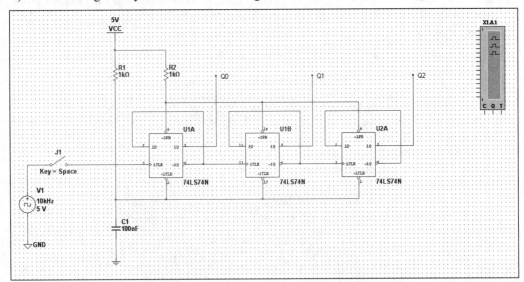

Figure 1-9: Circuit with Logic Analyzer

As Figure 1-9 shows, the logic analyzer is more complicated than a probe although it has the same basic function. It has 16 data inputs along its left side, each of which acts like a probe. Three control inputs on the bottom determine when the logic analyzer checks, or samples, the voltages on the data inputs. If the voltage is above the set threshold value (in this case the default value of 2.5 V), the logic analyzer records that the data sample is HIGH. If the voltage is below the threshold value, the logic analyzer records that the data sample is LOW.

When you specify that the logic analyzer should use an external clock, the logic analyzer uses the "C", or clock, input to synchronize when it takes samples.

The "Q", or qualifier, specifies whether an externally supplied clock should be HIGH (H), LOW (L), or either (X) to sample data.

The "T" input instructs the logic analyzer to look for a specific pattern of data inputs to begin sampling data.

4) Connect the first three data inputs on the top left side of the logic analyzer to the Q_0 through Q_2 junctions. Note that when you do so, the junctions will disappear.

5) Connect the logic analyzer's clock ("C"), qualifier ("Q"), and trigger ("T") inputs to a point on the right side of the switch. The circuit should now look like that in Figure 1-10.

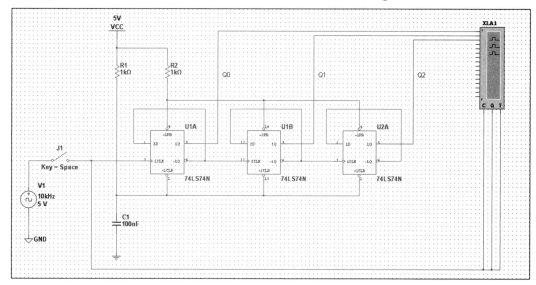

Figure 1-10: Circuit with Logic Analyzer Connections

6) Double-click on the logic analyzer to expand it. Figure 1-11 shows the expanded view.

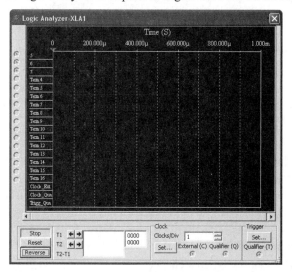

Figure 1-11: Logic Analyzer Expanded View

As the dots inside the connections for the data and control inputs show, the first three data lines and the external clock, qualifier, and trigger lines for the logic analyzer are connected.

7) Click the **Reverse** button to change the display background from black to white. This is optional, but may allow you to see the logic signals and measurement divisions more easily.

8) Set the **Clocks/Div** value in the **Clock** section to "1".

9) Click the **Set...** button in the **Clock** section. This will open the **Clock setup** window, shown in Figure 1-12.

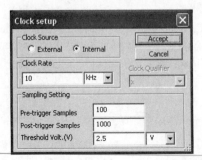

Figure 1-12: Logic Analyzer Clock Setup Window

10) Select **External** in the **Clock Source** section. This will set the logic analyzer to ignore the **Clock Rate** setting for its internal clock and to use the circuit clock, connected to the "C" input, to sample data.

11) Click the ▼ button next to the **Clock Qualifier** box and select "1" from the drop-down list.

12) Change the **Pre-trigger samples** value to "1". This will have the logic analyzer save the one data sample that occurs just before the specified trigger.

13) Change the **Post-trigger samples** value to "10". This will set the logic analyzer to collect ten samples. Because the logic analyzer is using the circuit clock to sample data, and is set for 1 clock per division, the logic analyzer will collect one data sample per division for ten clock pulses.

14) Click the **Accept** button to keep your changes.

15) Click the **Set...** button in the **Trigger** section. This will open the **Trigger Settings** window, shown in Figure 1-13.

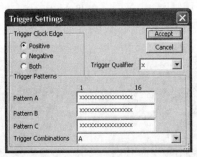

Figure 1-13: Trigger Settings Window

16) Select "Positive" in the **Trigger Clock Edge** section. This will cause the logic analyzer to look for a trigger when the input on its "T" input changes from LOW to HIGH.

17) Click the ▼ button next to the **Trigger Qualifier** box and select "1" from the drop-down list. This will require the "T" input to be HIGH when the logic analyzer finds a valid trigger.

18) Change the first three values in the **Pattern A** box from "xxx" to "000". This will instruct the logic analyzer to look for LOW signals on all three data inputs as the trigger qualifier.

19) Click the **Accept** button to keep your changes.

20) Start the simulation and allow the simulation to run for several seconds.

21) Press the space bar to close the switch J_1. This will connect the 10 kHz clock signal to the circuit.

22) When the logic analyzer stops collecting data, stop the simulation.

23) Side the scroll bar at the bottom of the logic analyzer display window all the way to the left to view the beginning of the waveform. The logic analyzer display should look similar to that in Figure 1-14.

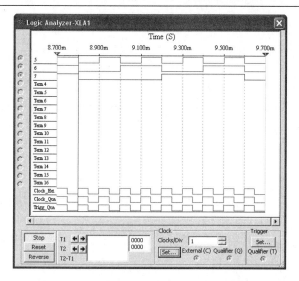

Figure 1-14: Logic Analyzer Data Display

The solid vertical line between the first and second divisions indicates when the logic analyzer found a valid trigger (000). The data to the left of the line is the pre-trigger data (data before a valid trigger occurred), and the data to the right is the post-trigger data (data after a valid trigger occurred).

24) Record in Table 1-3 the logic levels for each of the ten time divisions for the logic analyzer display. If the waveform for an output is low, write "LOW". If the waveform for an output is high, write "HIGH".

Table 1-3: Logic Analyzer Circuit States

Time Division	Q_0 (RED)	Q_1 (GREEN)	Q_2 (BLUE)
1			
2			
3			
4			
5			
6			
7			
8			
9			
10			

Conclusions for Part 3

Questions for Part 3

1) From Table 1-3, how many unique states does the circuit appear to have?

2) From Table 1-3, does there appear to be any pattern to how the state of each data line changes?

3) Does the circuit operation appear to be any different at 10 kHz compared to 10 Hz?

Digital Experiment 2 - Number Systems

2.1   Introduction

Signals in binary digital systems have only two valid states – LOW and HIGH. Because of this, the binary number system is a convenient way to represent digital data. Just as the decimal number system uses ten digits (0 through 9) to represent decimal values, the binary number system uses two digits (0 and 1) to represent binary values. These <u>bi</u>nary di<u>git</u>s, or *bits*, can represent the LOW and HIGH states of digital data. Figure 2-1 shows how a binary value can represent the states of a digital signal.

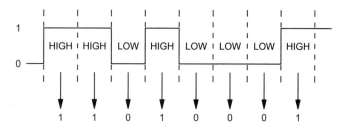

Figure 2-1: Representing a Digital Signal with a Binary Value

A problem with binary numbers is that people are not very good at working with long strings of zeroes and ones. Engineers, technicians, and programmers who work with computers and other digital devices must be able to convert binary values to a more familiar or convenient number system. In this experiment you will convert binary values to decimal, octal, and hexadecimal, and use Multisim circuits to verify that you have converted the numbers correctly. You will also compare the basic operation of combinational and sequential logic circuits.

The Multisim circuits for this experiment are provided so that you can observe and compare the operation of combinational and sequential logic circuits. At this point you do not need to know *how* they work. Note that the circuit in Part 1 will run very slowly, as it is more complex than the circuits in Parts 2 and 3 that you will use.

2.2   Reading

Floyd, *Digital Fundamentals, 10th Edition*, Chapter 2

2.3   Key Objectives

Part 1:   Apply the weighted-column method to convert binary values to corresponding decimal values and verify your answers with a sequential binary-to-decimal conversion circuit.

Part 2:   Apply the group-by-three method to convert binary values to corresponding octal values and verify your answers with a combinational binary-to-octal conversion circuit.

Part 3:   Apply the group-by-four method to convert binary value to corresponding hexadecimal values and verify your answers with a combinational binary-to-hexadecimal conversion circuit.

2.4   Multisim Circuits

Part 1:   *Digital_Exp_02_Part_01*

Part 2:   *Digital_Exp_02_Part_02*

Part 3:   *Digital_Exp_02_Part_03*

Part 1:   Binary-to-Decimal Conversion

A standard conversion technique for converting a binary value to a decimal value is the weighted-column method. This method multiplies the bit value for each column by the binary weight of that column and summing the result. Figure 2-2 illustrates this method.

COLUMN WEIGHT	$2^7 = 128$	$2^6 = 64$	$2^5 = 32$	$2^4 = 16$	$2^3 = 8$	$2^2 = 4$	$2^1 = 2$	$2^0 = 1$
COLUMN VALUE	1	0	1	0	0	1	1	0

SUM OF COLUMNS $128 + 0 + 32 + 0 + 0 + 4 + 2 + 0 = 166$

Figure 2-2: Weighted-Column Conversion Method

Another technique, used in simple computer programs, is to count down in binary to 0 while simultaneously counting up from 0 in decimal. When the binary count equals 0, the decimal value equals the original binary value.

In this part of the experiment you will use the weighted-column method to convert binary values to decimal and verify your answers with a Multisim circuit that uses the simultaneous counting method.

1) For each of the binary values in Table 2-1, enter the weighted-column value for the bit that corresponds to that column.

- If the bit that corresponds to a column is "0", enter "0".
- If the bit that corresponds to a column is "1", enter the indicated weight for that column.

2) For each binary value in Table 2-1, sum the column weight values and enter the sum in the "Calculated Value" column for that row.

Table 2-1: Weighted-Column Binary Conversion

Binary Value	Column Weights								Calculated Value
	128	64	32	16	8	4	2	1	
00000001									
00010000									
01100100									
10101010									
11000011									
11111111									

3) Open the file *Digital_Exp_02_Part_01*.

4) Set the DIP switch J_1 to the first binary value in Table 2-2.

- Click on a switch to slide the white portion down for a "0".
- Click on a switch to slide the white portion up for a "1".

5) Start the simulation.

6) Press the "L" key to close switch J_2. This will clear the decimal display and load the binary value into the binary counter.

7) Verify that the probes X_1 through X_8 match the binary value. A lit probe indicates a "1", and an unlit probe indicates a "0".

8) Press the "L" key to open switch J_2. This will begin the conversion and light up probe X_9 to indicate that the conversion is in progress. The decimal value will change as the counters sequence through the count.

9) When probe X_9 goes off, indicating that the conversion is complete, stop the simulation.

10) Copy the value in the digital display into the "Computed Value" column of Table 2-2.

11) Repeat steps 4 through 10 for the remaining binary values in Table 2-2.

Table 2-2: Multisim Computed Decimal Values

Binary Value	Computed Value	Binary Value	Computed Value
00000001		10101010	
00010000		11000011	
01100100		11111111	

Observations:

Questions for Part 1

1) How does the simulation time (time for the circuit to compute the decimal value) change as the binary values get larger? Why?

2) Set the DIP switch J_1 to 11111111, start the conversion process, and watch the simulation time at the lower right-hand corner of the Multisim screen. How does the simulation time (the time the circuit experiences) compare to real time (the time you experience)?

Part 2: Binary-to Octal Conversion

Converting binary values to values in binary-based number systems such as Base 8 (octal) and Base 16 (hexadecimal) is much simpler than converting binary values to non-related number systems such as Base 10 (decimal). Because eight is the third power of two, converting a binary value to octal consists simply of grouping bits by three, beginning from the right, and then converting the groups to octal digits. Figure 2-3 illustrates this method. Note that the leftmost group may, as in this case, contain less than three bits. When this happens, assume the missing bits in the group are zeros.

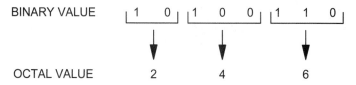

Figure 2-3: Group-by-Three Conversion Method

In this part, you will use the group-by-three method to convert binary values to octal and use a combinational Multisim circuit that groups bits to verify your answer. Table 2-3 shows the 3-bit groups that correspond to each octal digit.

1) For the first binary value in Table 2-4, write the bit groups in the corresponding "Bit Group" column of Table 2-4. Be sure to start from the right when grouping by three!

2) Use Table 2-3 to record the octal value of each bit group in the corresponding "Octal Value" column of Table 2-4.

Table 2-3: Octal Digits for 3-Bit Binary Groups

Bit Group	000	001	010	011	100	101	110	111
Octal Digit	0	1	2	3	4	5	6	7

3) Repeat steps 1 and 2 for the remaining binary values in Table 2-4.

Table 2-4: Group-by-Three Conversion Method

Binary Value	Bit Group			Octal Value		
	Left	Middle	Right	Digit 1	Digit 2	Digit 3
00000011						
00001110						
01010101						
10000001						
10100110						
11111111						

4) Open the file *Digital_Exp_02_Part_02*.

5) Start the simulation.

6) Set the DIP switch J_1 to the first binary value in Table 2-5.

   - Click on a switch to slide the white portion down for a "0".
   - Click on a switch to slide the white portion up for a "1".

7) Verify that the probes X_1 through X_8 match the binary value. A lit probe indicates a "1", and an unlit probe indicates a "0".

8) Copy the value in the digital display into the "Computed Value" column of Table 2-5.

9) Repeat steps 6 through 8 for the remaining binary values in Table 2-5.

10) Stop the simulation.

Table 2-5: Multisim Computed Octal Values

Binary Value	Computed Value	Binary Value	Computed Value	Binary Value	Computed Value
00000011		01010101		10100110	
00001110		10000001		11111111	

Observations:

Questions for Part 2

1) How does the simulation time (time for the circuit to compute the decimal value) change as the binary values get larger?

2) Compare the display circuitry for *Digital_Exp_02_Part_01* and *Digital_Exp_02_Part_02*. Note that the binary-to-octal conversion circuit uses the same display circuitry as the binary-to-decimal conversion circuit. Why is this possible?

Part 3:   Binary-to-Hexadecimal Conversion

Binary-to-hexadecimal conversion is similar to binary-to-octal conversion, except for the number of bits in each group. Because sixteen is the fourth power of two, converting a binary value to hexadecimal consists of grouping bits by four, beginning from the right, and converting each group to a hexadecimal digit. Figure 2-4 illustrates this method. If the leftmost group contains fewer than four bits, assume the missing bits are all zeros.

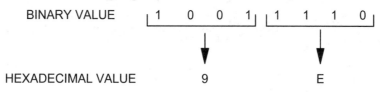

Figure 2-4: Group-by-Four Conversion Method

Hexadecimal (or "hex") is a popular number format in computing, as computers work with binary values that are multiples of four bits. In this part, you will use the group-by-four method to convert binary values to hexadecimal and use a combinational Multisim circuit that groups bits to verify your answer. Table 2-6 shows the 4-bit groups that correspond to each hexadecimal digit.

1) For the first binary value in Table 2-7, write the bit groups in the corresponding "Bit Group" column of Table 2-7. Be sure to start from the right when grouping by four!

2) Use Table 2-6 to record the hexadecimal value of each bit group in the corresponding "Hexadecimal Value" column of Table 2-7.

Table 2-6: Hexadecimal Digits for 4-Bit Binary Groups

Bit Group	0000	0001	0010	0011	0100	0101	0110	0111
Hexadecimal Digit	0	1	2	3	4	5	6	7
Bit Group	1000	1001	1010	1011	1100	1101	1110	1111
Hexadecimal Digit	8	9	A	B	C	D	E	F

3) Repeat steps 1 and 2 for the remaining binary values in Table 2-7.

Table 2-7: Group-by-Four Conversion Method

Binary Value	Bit Group		Hexadecimal Value	
	Left	Right	Digit 1	Digit 2
00010001				
01111110				
10101111				
11000010				
11100101				
11111111				

4) Open the Multisim file *Digital_Exp_02_Part_03*.

5) Start the simulation.

6) Set the DIP switch J_1 to the first binary value in Table 2-8.

   - Click on a switch to slide the white portion down for a "0".
   - Click on a switch to slide the white portion up for a "1".

7) Verify that the probes X_1 through X_8 match the binary value. A lit probe indicates a "1", and an unlit probe indicates a "0".

8) Copy the value in the digital display into the "Computed Value" column of Table 2-8.

9) Repeat steps 6 through 8 for the remaining binary values in Table 2-8.

10) Stop the simulation.

Table 2-8: Multisim Computed Hexadecimal Values

Binary Value	Computed Value	Binary Value	Computed Value
00010001		11000010	
01111110		11100101	
10101111		11111111	

Observations:

Questions for Part 3

1) What are some advantages of converting values from binary format to hexadecimal format compared to converting values from binary format values to octal format?

2) Why does the binary-to-hexadecimal converter circuit use different display circuitry than the binary-to-decimal display circuit?

Name _____   Class _____

Date _____   Instructor_____

Digital Experiment 3 - Logic Gates

3.1   Introduction

All combinational logic can be reduced to basic AND, OR, and NOT logic operations and the AND gate, OR gate, and inverter that implement them. Understanding the operation of these basic logic gates and recognizing how these combine to create more complex logic are the foundation of digital electronics. This experiment will demonstrate how to analyze the operation of logic gates and verify their operation. In Part 1 of this experiment, you will use Multisim to verify the operation of the AND gate, OR gate, and inverter. In Part 2, you will use the Multisim logic converter to examine the operation of the NAND, NOR, and XOR gates.

3.2   Reading

Floyd, *Digital Fundamentals, 10th Edition*, Chapter 3.

3.3   Key Objectives

Part 1:   Learn how to use switches and probes to verify the operation of the AND gate, OR gate, and inverter.

Part 2:   Learn how to use the logic converter to verify the operation of the NAND, NOR, and XOR (exclusive-OR) gates.

3.4   Multisim Circuits

Part 1:   *Digital_Exp_03_Part_01a, Digital_Exp_03_Part_01b*, and *Digital_Exp_03_Part_01c*.

Part 2:   *Digital_Exp_03_Part_02a, Digital_Exp_03_Part_02b*, and *Digital_Exp_03_Part_02c*.

Part 1:   Basic Logic Gates

3.5   The AND Gate

In this section you will connect components to the AND gate so that you can determine its output for each combination of inputs (i.e., its truth table).

1) Open the file *Digital_Exp_03_Part_01a*.

2) Select the **Place Basic** tool from the **Component Toolbar**.

3) Select "SWITCH" from the **Family:** window and "SPST" from the **Component:** list.

4) Left-click to the left and a little above the AND gate to place the switch. Note that Multisim automatically numbers the reference designator as "J1".

5) Right-click on J_1 and select "Properties" from the right-click menu.

6) Click the down arrow (▼) button to the right of the **Key for Switch** window and select "A" from the drop-down list. This will configure the "A" key to open and close the switch.

7) Place another SPST switch to the left and a little below the AND gate. Note that Multisim automatically numbers the reference designator as "J2".

8) Right-click on J_2 and select "Properties" from the right-click menu.

9) Click the down arrow (▼) button to the right of the **Key for Switch** window and select "B" from the drop-down list. This will configure the "B" key to open and close the switch.

10) Connect the right side of J_1 to the top input of the AND gate.

11) Connect the right side of J_2 to the bottom input of the AND gate. Your circuit should now look like Figure 3-1.

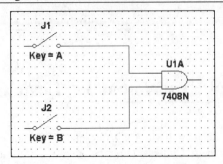

Figure 3-1: Circuit with Switches Placed

12) Select the **Place Basic** tool from the **Component Toolbar**.

13) Select "RESISTOR" from the **Family:** window and "1k" from the **Component:** list.

14) Left-click above and between J_1 and the AND gate to place the 1 kΩ resistor. Note that Multisim automatically numbers the reference designator as "R1".

15) Right-click on R_1 and select "90 clockwise" from the right-click menu. This will rotate the resistor so that its orientation is vertical.

16) Place and rotate another 1 kΩ resistor to the right of R_1. Note that Multisim automatically renumbers the reference designator as "R2".

17) Connect the bottom of R_1 to the wire between J_1 and the AND gate.

18) Connect the bottom of R_2 to the wire between J_2 and the AND gate. Your circuit should now look something like Figure 3-2.

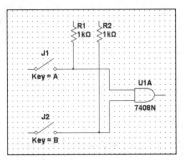

Figure 3-2: Circuit with Resistors Placed

19) Select the **Place Source** tool from the **Component Toolbar**.

20) Select "Power Sources" from the **Family:** window and "VCC" from the **Component:** list.

21) Left-click above R_1 and R_2 to place the source.

22) Select the **Place Source** tool from the **Component Toolbar**.

23) Select "Power Sources" from the **Family:** window and "DGND" from the **Component:** list.

24) Left-click to the left and below J_1 and J_2 to place the digital ground.

25) Connect the tops of both R_1 and R_2 to the VCC source.

26) Connect the left sides of both J_1 and J_2 to the digital ground.

27) Select the **Place Indicator** tool from the **Component Toolbar**.

28) Select "PROBE" from the **Family:** window and "PROBE_DIG_RED" from the **Component:** list.

29) Left-click above and to the right of the AND gate to place the probe.

30) Connect the probe to the output of the AND gate. Your circuit should now look like Figure 3-3.

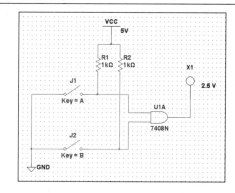

Figure 3-3: Completed Test Circuit

You are now ready to test the circuit. When a switch is closed, its input is pulled down to digital ground through the switch to a logic "0". When a switch is open, its input is pulled up to VCC through the resistor to a logic "1". When the probe is not lit, the output is LOW, or a logic "0". When the probe is lit, the output is HIGH, or a logic "1".

31) Click the **Run** switch to start the simulation.

32) Use the "A" and "B" keys to open and close the switches. Record the state of the probe for each combination of switch settings in the "Output" column of Table 3-1.

33) Stop the simulation.

Table 3-1: AND Gate Truth Table

Switch A Closed = 0 Open = 1	Switch B Closed = 0 Open = 1	Output Not Lit = 0 Lit = 1
0	0	
0	1	
1	0	
1	1	

3.6 The OR Gate

In this section you will determine the truth table for the OR gate.

1) Open the file *Digital_Exp_03_Part_01b*.

2) Use the procedure from Section 3.5 to connect the test circuitry to the OR gate.

3) Use the "A" and "B" keys to open and close the switches. Record the state of the probe for each combination of switch settings in the "Output" column of Table 3-2.

4) Stop the simulation.

Table 3-2: OR Gate Truth Table

Switch A Closed = 0 Open = 1	Switch B Closed = 0 Open = 1	Output Not Lit = 0 Lit = 1
0	0	
0	1	
1	0	
1	1	

3.7   The Inverter

In this section you will determine the truth table for the inverter.

1) Open the file *Digital_Exp_03_Part_01c*.

2) Use the procedure from Section 3.5 to connect the test circuitry to the inverter. Note that because the inverter has only one input, the test circuit requires only one switch and one resistor. Use the **Key for Switch** window to configure the "A" key to open and close the switch.

3) Use the "A" key to open and close the switches. Record the state of the probe (0 for not lit and 1 for lit) for each combination of switch settings in the "Output" column of Table 3-3.

4) Stop the simulation.

Table 3-3: Inverter Truth Table

Switch A Closed = 0 Open = 1	Output Not Lit = 0 Lit = 1
0	
1	

Observations:

Conclusions for Part 1

Questions for Part 1

1) Which gate follows the rule "Any 1 gives a 1"?

2) Which gate follows the rule "Any 0 gives a 0"?

3) Explain why the inverter is so-named.

Part 2:   Extended Logic Gates

Extended logic gates are similar to basic logic gates in that they implement useful and well-defined logic functions, but their logic functions can be implemented using basic AND gates, OR gates, and inverters. The most common of these are NAND, NOR, and XOR gates. In this part of the experiment, you will use the Multisim logic converter to determine the truth tables for these gates. The logic converter is not a real laboratory device, although its function can be simulated using real digital instruments and software. Figure 3-4 shows the **Logic Converter** tool on the **Instruments Toolbar**. Figure 3-5 shows the minimized and expanded views of the logic converter.

Figure 3-4: Logic Converter Tool

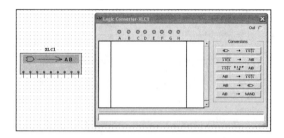

Figure 3-5: Minimized and Expanded Views of Logic Converter

3.8 The NAND Gate

As its name suggests, the NAND (for "not AND") gate combines the logic of an AND gate and an inverter.

1) Open the file *Digital_Exp_03_Part_02a*.

2) Select the **Logic Converter** tool from the **Instruments Toolbar**.

3) Left-click above the NAND gate to place the logic converter.

4) Connect the "A" (far left) terminal of the logic converter to the top input of the NAND gate.

5) Connect the "B" (second from far left) terminal of the logic converter to the bottom input of the NAND gate.

6) Connect the output of the NAND gate to the "Out" (far right) terminal of the logic converter.

7) Double-click on the logic converter to expand it. Your circuit should look like Figure 3-6.

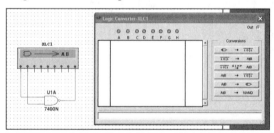

Figure 3-6: Completed Test Circuit

8) Click the top button in the "Conversions" section. This will convert the circuit representation of the NAND gate to a truth table representation. Note that you do not have to click the "Run" button.

9) Copy the logic converter results into the corresponding columns of Table 3-4. The unlabelled column at the far right of the logic converter is the output of the NAND gate.

Table 3-4: NAND Gate Truth Table

A	B	Output
0	0	
0	1	
1	0	
1	1	

3.9   The NOR Gate

As its name suggests, the NOR (for "not OR") gate combines the logic of an OR gate and an inverter.

1) Open the file *Digital_Exp_03_Part_02b*.

2) Use the logic converter to generate the truth table for the NOR gate and copy the results into Table 3-5.

Table 3-5: NOR Gate Truth Table

A	B	Output
0	0	
0	1	
1	0	
1	1	

3.10 The XOR Gate

The exclusive-OR (XOR) gate is similar to the OR gate in that its output is 1 when either input is 1.

1) Open the file *Digital_Exp_03_Part_02c*.

2) Use the logic converter to generate the truth table for the XOR gate and copy the results into Table 3-6.

Table 3-6: XOR Gate Truth Table

A	B	Output
0	0	
0	1	
1	0	
1	1	

Observations:

Conclusions for Part 2

Questions for Part 2

1) How does the truth table of the NAND gate compare with the truth table of the AND gate?

2) You observe that the output of an unknown 2-input gate is 0 only when the inputs are both 0 or both inputs are 1. What type of logic gate are you observing?

3) What gate follows the rule "Any 1 gives a 0"?

4) An XOR gate is similar to an OR gate in that the output is 1 when either input is 1. How does XOR gate differ from an OR gate?

Digital Experiment 4 - Boolean Algebra and DeMorgan's Theorem

4.1   Introduction

A logic diagram or schematic drawing, such as those you create in Multisim, is one way to represent logic functions. A truth table, which shows which combinations of inputs produce a 0 or 1, is another way. Boolean algebra is a third way. Boolean algebra was developed by George Boole in the mid-1800s as a means of representing and analyzing logic expressions, and uses variables and operators, much like mathematical equations, to represent logic expressions. Boolean algebra typically represents the AND function as a product, the OR function as a sum, and the invert function with an overbar over the inverted term. Figure 4-1 shows the Boolean operators for the basic logic functions. Note that, because some software does not support overbars, a forward slash (/) before a term or an apostrophe (') after a term can also represent the invert function. Multisim uses an apostrophe after the inverted term, although some Multisim parts use a tilde (~) before a signal name.

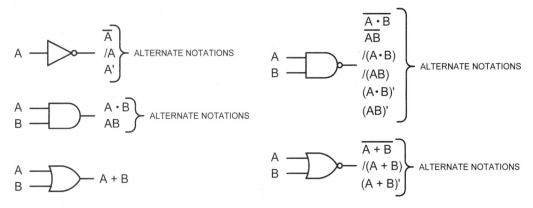

Figure 4-1: Boolean Expressions for Basic Logic Gates

Boolean operators, like mathematical operators, have a standard precedence, or order of operation. The invert operator has the highest precedence and the AND operator takes precedence over the OR operator, but parentheses alter the precedence of these operators. The expressions for the NAND and NOR gates, for example, require parentheses so that the entire expression, rather than just one variable, is inverted.

You can use Boolean algebra to simplify logic expressions, as well as represent and evaluate logic functions. Boolean algebra has 12 basic rules that summarize functionally equivalent logic expressions. These rules allow you to work through and simplify more complex Boolean expressions. In Part 1 of this experiment, you will use Multisim to verify these rules.

Two additional theorems for working with Boolean expressions are DeMorgan's theorems. These theorems mathematically verify that NAND gates are equivalent to negative-OR gates, and that NOR gates are equivalent to negative-AND gates. The theorems are useful for simplifying expressions with inverted inputs and outputs and working with negative logic, in which 0 rather than 1 is the active logic level. In Part 2 of this experiment, you will use Multisim to verify DeMorgan's theorems.

4.2   Reading

Floyd, *Digital Fundamentals, 10th Edition*, Chapter 4

4.3   Key Objectives

Part 1:   Verify the validity of the 12 rules of Boolean algebra.

Part 2:   Verify the validity of DeMorgan's theorems.

4.4 Multisim Files

Part 1: *Digital_Exp_04_Part_01a* through *Digital_Exp_04_Part_01q*.

Part 2: *Digital_Exp_04_Part_02a* and *Digital_Exp_04_Part_02b*.

Part 1: Boolean Algebra

4.5 Laws of Boolean Algebra

For each of the Multisim files listed in Table 4-1:

1) Open the file.

2) Verify that the logic driving each probe corresponds to the indicated logic expression.

3) Run the simulation.

4) Set the circuit switches to each setting listed in the circuit file.

5) Specify in the table whether the states for each probe in the circuit are identical for all switch positions.

6) Stop the simulation.

Table 4-1: Verification of Laws of Boolean Algebra

Circuit File	Represented Law	Probe States Identical for All Switch Positions?
Digital_Exp_04_01a	$A + B = B + A$	
Digital_Exp_04_01b	$A \cdot B = B \cdot A$	
Digital_Exp_04_01c	$A + (B + C) = (A + B) + C$	
Digital_Exp_04_01d	$A \cdot (B \cdot C) = (A \cdot B) \cdot C$	
Digital_Exp_04_01e	$A \cdot (B + C) = A \cdot B + A \cdot C$	

4.6 Rules of Boolean Algebra

For each of the Multisim files listed in Table 4-2:

1) Open the file.

2) Verify that the logic driving each probe corresponds to the indicated logic expression.

3) Run the simulation.

4) Set the circuit switches to each setting listed in the circuit file.

5) Specify in the table whether the states for each probe in the circuit are identical for all switch positions.

6) Stop the simulation.

Table 4-2: Verification of Rules of Boolean Algebra (continued on next page)

Circuit File	Represented Rule	Probe States Identical for All Switch Positions?
Digital_Exp_04_01f	$A + 0 = A$	
Digital_Exp_04_01g	$A + 1 = 1$	
Digital_Exp_04_01h	$A \cdot 0 = 0$	
Digital_Exp_04_01i	$A \cdot 1 = A$	
Digital_Exp_04_01j	$A + A = A$	

Circuit File	Represented Rule	Probe States Identical for All Switch Positions?
Digital_Exp_04_01k	$A + A' = 1$	
Digital_Exp_04_01l	$A \cdot A = A$	
Digital_Exp_04_01m	$A \cdot A' = 0$	
Digital_Exp_04_01n	$(A')' = A$	
Digital_Exp_04_01o	$A + AB = A$	
Digital_Exp_04_01p	$A + A'B = A$	
Digital_Exp_04_01q	$(A + B)(A + C) = A + BC$	

Conclusions for Part 1

Questions for Part 1

1) What principle was used to verify the laws and rules of Boolean algebra?

2) Using the rules of Boolean algebra, what is the simplified expression for $AA + AB + AC$?

3) From Rules 3 and 4, how can a 2-input AND gate be used to selectively pass or block a signal?

Part 2:   DeMorgan's Theorems

4.7   NAND Gate Equivalent

1) Open the Multisim file *Digital_Exp_04_Part_02a*.
2) Run the simulation.
3) Apply the switch inputs shown in Table 4-3. Record the states (unlit = 0, lit = 1) of the probe for $(AB)'$ and the probe for $A' + B'$ for each switch setting.

Table 4-3: Results of NAND Equivalent Switch Settings

Switch *B*	Switch *A*	Probe $(AB)'$	Probe $A' + B'$
0	0		
0	1		
1	0		
1	1		

4) Stop the simulation and summarize your observations.

4.8   NOR Gate Equivalent

1) Open the Multisim file *Digital_Exp_04_Part_02b*.

2) Run the simulation.

3) Apply the switch inputs shown in Table 4-4. Record the states (unlit = 0, lit = 1) of the probe for $(A + B)$' and the probe for $A'B'$ for each switch setting.

Table 4-4: Results of NOR Equivalent Switch Settings

Switch B	Switch A	Probe $(A + B)$'	Probe $A'B'$
0	0		
0	1		
1	0		
1	1		

4) Stop the simulation and summarize your observations.

Conclusions for Part 2

Questions for Part 2

1) Using DeMorgan's Theorems, what is an equivalent expression for $((A + B)'(C + D)')$'?

2) You wish to implement a circuit that will output $X = 1$ if either input A or B is 0. What gate should you use?

Digital Experiment 5 - Combinational Logic

5.1 Introduction

Individual logic gates are limited in what they can do. However, circuits that consist of multiple gates can make decisions, work through mathematical computations, control other circuits, and perform many other complex operations. Even the most advanced and complicated microprocessors consist of many interconnected basic gates. Combinational logic circuits are circuits that combine logic gates so that the state of the inputs alone determines the state of the output. The binary-to-octal and binary-to-hexadecimal converters in Experiment 2 are examples of combinational logic circuits. The value on the output display depended only upon the state of the input switches.

In Part 1 of this experiment, you will examine how to design and implement a 2-of-3 voting circuit by 1) defining the relationship between the input and output states with a truth table, 2) translating the truth table entries into a Boolean expression, and 3) verifying the circuit implementation for the Boolean expression by simulating the circuit in Multisim. In Part 2, you will examine how to simplify combinational logic using Boolean algebra and a graphical tool called a *Karnaugh map* and verify the simplified expression for the 2-of-3 voting circuit.

5.2 Reading

Floyd, *Digital Fundamentals, 10th Edition*, Chapters 4 and 5

5.3 Key Objectives

Part 1:   Analyze, define, and verify the logic for a 2-of-3 voting circuit.

Part 2:   Use Boolean algebra and Karnaugh map techniques to simplify the logic for the 2-of-3 voting circuit.

5.4 Multisim Files

Part 1:   *Digital_Exp_05_Part_01*

Part 2:   *Digital_Exp_05_Part_02*

Part 1:   The 2-of-3 Voting Circuit

A panel of three judges for the *Digital Idol* circuit design competition needs a circuit that will indicate whether a digital circuit project will progress to the next level of the competition. A project will advance to the next level if at least two of the judges approve the project with a "Yes" vote. A project will not advance if at least two of the judges reject the project with a "No" vote.

5.5 Defining the Logic Function

Complete the truth table in Table 5-1 by indicating whether or not each combination of "No" and "Yes" votes cast by the three judges (called A, B, and C) will approve the project so that it advances to the next level of competition.

Table 5-1: Truth Table for Voting Logic (0 = "No", 1 = "Yes")

JUDGE			PROJECT APPROVED	JUDGE			PROJECT APPROVED
A	B	C		A	B	C	
0	0	0	0	1	0	0	
0	0	1	0	1	0	1	
0	1	0	0	1	1	0	
0	1	1	1	1	1	1	

5.6   Translating the Truth Table

Translating the truth table into a Boolean expression consists of ANDing the input terms for each truth table entry for which the output is "1", and then ORing these product terms together to create a sum-of-product (SOP) expression for the logic function. If the input term for an entry is a "1" use the true (noninverted) form of the input. If the input term for an entry is a "0", use the complement (inverted) form.

Example: What AND term corresponds to $A = 0$, $B = 1$, and $C = 1$?

The A variable is "0" so the complement A' is used. The B variable is "1" so the true form B is used. The C variable is "1", so the true form C is used. The corresponding AND term is $A'BC$.

Use the data in Table 5-1 to write the SOP expression for 2-of-3 voting logic function.

5.7   Verifying the 2-of-3 Voting Logic

1) Open the Multisim file *Digital_Exp_05_Part_01*.

2) For each combination of votes in Table 5-2,

   a.   Use the "A", "B", and "C" keys to set the vote for each judge. Alternatively, you can click on each switch to open or close it.

   b.   Record the result of the vote Table 5-2.

3) Compare the results of Table 5-2 with the results of Table 5-1. Do the table results match?

Table 5-2: Results of Voting Logic Circuit

SWITCH SETTINGS (Closed = 0, Open = 1)			OUTPUT STATE (OFF = 0, ON = 1)	SWITCH SETTINGS (Closed = 0, Open = 1)			OUTPUT STATE (OFF = 0, ON = 1)
A	*B*	*C*		*A*	*B*	*C*	
0	0	0		1	0	0	
0	0	1		1	0	1	
0	1	0		1	1	0	
0	1	1		1	1	1	

Conclusions for Part 1

Questions for Part 1

1) Why does the Multisim circuit *Digital_Exp_05_Part_01* use three cascaded OR gates to OR the AND terms together?

2) Suppose you wish to modify the circuit in *Digital_Exp_05_Part_01* so that the output indicates that a project is rejected rather than approved. How could you do this?

Part 2: Logic Simplification

5.8 Boolean Algebra

The laws and rules of Boolean algebra shown in Table 5-3 can be used to simplify logic expressions so that circuits require fewer gates with fewer inputs. This reduces circuit complexity and the costs of parts and manufacturing. Advantages of Boolean algebra are that it rigorously proves that two logic expressions are equivalent and it can be used to reduce expressions with (in theory) any number of variables and terms. A disadvantage is that Boolean algebra does not guarantee that an expression is the minimum SOP expression or even that you will find the minimum SOP expression. Another disadvantage is that there is no "standard procedure" on how to proceed, so that some ingenuity is often required in simplifying an expression and more than one process can arrive at the same result.

Table 5-3: Summary of Rules and Laws of Boolean Algebra

Reference	Rule or Law	Reference	Rule or Law
Rule 1	$A + 0 = A$	Rule 10	$A + AB = A$
Rule 2	$A + 1 = 1$	Rule 11	$A + A'B = A$
Rule 3	$A \cdot 0 = 0$	Rule 12	$(A + B)(A + C) = A + BC$
Rule 4	$A \cdot 1 = A$	Commutative Law	$A + B = B + A$
Rule 5	$A + A = A$		$AB = BA$
Rule 6	$A + A' = 1$	Associative Law	$A + (B + C) = (A + B) + C$
Rule 7	$A \cdot A = A$		$A(BC) = (AB)C$
Rule 8	$A \cdot A' = 0$	Distributive Law	$A(B + C) = AB + AC$
Rule 9	$(A')' = A$		

Table 5-4 shows a sequence that simplifies the 2-of-3 voting to its minimum SOP form, but you must determine the rule or law that validates each step.

1) Begin with the Boolean expression in the first row of Table 5-4.

2) Indicate the Boolean rule or law in Table 5-3 that is applied to derive the new expression for each row of Table 5-4 (the first two simplifications are done for you)

3) Continue through the table, indicating the Boolean rule or law that is applied at each step to simplify the 2-of-3 voting logic to its minimum SOP form.

Table 5-4: Boolean Algebra Simplification of 2-of-3 Voting Logic

Step	Boolean Expression	Boolean Rule or Law	Step	Boolean Expression	Boolean Rule or Law
1	$A'BC + AB'C + ABC' + ABC$	Starting form	8	$A'BC + AB + AC$	
2	$A'BC + AB'C + AB(C' + C)$	Distributive Law	9	$(A'C + A)B + AC$	
3	$A'BC + AB'C + AB \cdot 1$	Rule 6	10	$(A + A'C)B + AC$	
4	$A'BC + AB'C + AB$		11	$(A + C)B + AC$	
5	$A'BC + A(B'C + B)$		12	$AB + CB + AC$	
6	$A'BC + A(B + B'C)$		13	$AB + BC + AC$	
7	$A'BC + A(B + C)$				

What does the final logic expression indicate?

5.9   Karnaugh Map

The Karnaugh map is a structured way to simplify a Boolean expression. Its main advantage is that it can always reduce an expression to its simplest SOP form. Its major drawback is that it is limited to simplifying expressions with five or fewer input variables.

1) Use the data from Table 5-1 to fill each cell of the blank Karnaugh map in Figure 5-1 with the output corresponding to the combination of input variables represented by that cell.

2) Group the cells containing a "1" according to the following rules:

   a.   Each cell containing a "1" must be included in at least one group. Cells may, but need not be, included in more than one group.

   b.   Cells in each group must be horizontally or vertically adjacent to each other (i.e., no diagonal groups). Karnaugh maps "wrap around" so that the top and bottom rows are vertically adjacent to each other and the leftmost and rightmost columns are horizontally adjacent to each other.

   c.   Groups must be essentially rectangular in form, so that every row in a group includes the same number of columns, and every column in a group includes the same number of rows.

   d.   The number of cells in each group must be a power of 2, so that each group contains 1 cell, 2 cells, 4 cells, etc.

   e.   Each group must contain as many cells as possible.

A B \ C	0	1
0 0	A' B' C'	A' B' C
0 1	A' B C'	A' B C
1 1	A B C'	A B C
1 0	A B' C'	A B' C

A B \ C	0	1
0 0		
0 1		
1 1		
1 0		

Figure 5-1: Three-Variable Karnaugh Map

3) Each group corresponds to an AND term in the final simplified expression. To determine the variables AND term, write down the input variables that are common to each cell in the group.

   AND Term 1:

   AND Term 2:

   AND Term 3:

4) OR the AND terms together to derive the final expression.

   Final expression:

5.10   Verification of Simplified Logic

1) Open the Multisim file *Digital_Exp_05_Part_02*.

2) For each combination of votes,

   a.   Use the "A", "B", and "C" keys to set the vote for each judge. Alternatively, you can click on each switch to open or close it.

   b.   Record the result of the vote Table 5-5.

3) Compare the results of Table 5-5 with the results of Table 5-2. Do the table results match?

Table 5-5: Results of Simplified Voting Logic Circuit

SWITCH SETTINGS (Closed = "0", Open = "1")			OUTPUT STATE (OFF = 0, ON = 1)	SWITCH SETTINGS (Closed = "0", Open = "1")			OUTPUT STATE (OFF = 0, ON = 1)
A	*B*	*C*		*A*	*B*	*C*	
0	0	0		1	0	0	
0	0	1		1	0	1	
0	1	0		1	1	0	
0	1	1		1	1	1	

Conclusions for Part 2

Questions for Part 2

1) How does simulating the Multisim circuit *Digital_Exp_05_Part_02* verify that the simplified logic expression is functionally equivalent to the original expression?

2) If you wished to ensure that a three-variable Boolean expression was in its minimum SOP form, which of the two simplification methods would you use? Why?

3) Is the Karnaugh map group for the expression $A'B'C' + A'B'C + A'BC' + ABC'$ shown in Figure 5-2 a valid group? Why or why not?

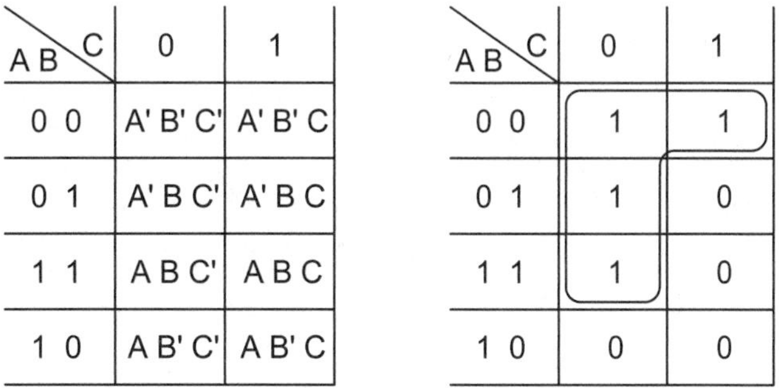

A B \ C	0	1
0 0	A' B' C'	A' B' C
0 1	A' B C'	A' B C
1 1	A B C'	A B C
1 0	A B' C'	A B' C

A B \ C	0	1
0 0	1	1
0 1	1	0
1 1	1	0
1 0	0	0

Figure 5-2: Proposed Karnaugh Map Group

Name _____   Class _____

Date _____   Instructor _____

Digital Experiment 6 - The Full Adder

6.1   Introduction

One class of combinational logic consists of decision circuits, in which the circuit evaluates the input signals to determine which specific action to take. The 2-of-3 voting logic circuit in Experiment 5 is an example of a decision circuit. Another important class of combinational logic consists of arithmetic functions, in which the circuit processes numeric (binary) data. Many of the earliest digital circuits were designed to perform high-speed calculations, and arithmetic functions today are found in a wide variety of calculating and measuring devices. The basis of arithmetic functions is a device called the *adder*, which, as its name suggests, adds two bits. There are two basic varieties of adders, called the *half-adder* and *full adder*.

The half-adder has two inputs, corresponding to two input bits, and two outputs, which indicate the *sum* and *carry* that result from adding the two input bits. A limitation of the half-adder is that it cannot add groups of bits, as the carry from one column cannot propagate to the next column. The full adder is similar to the half-adder but adds a third input, called the *carry in*. The carry in allows the full adder to accept the carry bit from a previous adder so that groups of bits can be added. Figure 6-1 shows the diagram of a half-adder and full adder.

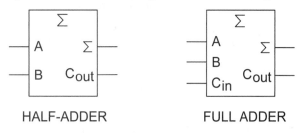

Figure 6-1: Diagrams of Half-Adder and Full Adder

In Part 1 of this experiment, you will examine and verify the operation of a full adder circuit. In Part 2, you will examine how a 4-bit full adder is cascaded to add 8-bit and wider numbers.

6.2   Reading

Floyd, *Digital Fundamentals, 10th Edition*, Chapter 6.

6.3   Key Objectives

Part 1:   Examine, verify, and compare the gate-level implementations of the full adder.

Part 2:   Examine and verify the operation of cascaded 4-bit binary adders.

6.4   Multisim Files

Part 1:   *Digital_Exp_06_Part_01a* and *Digital_Exp_06_Part_01b*

Part 2:   *Digital_Exp_06_Part_02a*, *Digital_Exp_06_Part_02b*, and *Digital_Exp_06_Part_02c*

Part 1:   The Full-Adder

6.5   Implementation 1

1) For each row in Table 6-1, record the expected sum and carry values for the "C_{in}", "A" and "B" inputs for a full adder in the "Σ" and "C_{out}" columns.

2) Open the Multisim file *Digital_Exp_06_Part_01a*.

3) Start the simulation.

4) Use the "A", "B", and "C" keys to open and close the "Input A", "Input B", and "Carry In" switches for each of the settings shown in Table 6-2. Alternatively, you can click on the switches to open or close them.

5) Record the state of the "Sum" and "Carry Out" probes for each combination of switch settings in the "Σ" and "C$_{out}$" columns of Table 6-2.

6) Stop the simulation.

Table 6-1: Sum and Carry Outputs for Full Adder

C$_{in}$	A	B	Σ	C$_{out}$	C$_{in}$	A	B	Σ	C$_{out}$
0	0	0			1	0	0		
0	0	1			1	0	1		
0	1	0			1	1	0		
0	1	1			1	1	1		

Table 6-2: Observed Outputs for Full Adder Implementation 1

Carry In	Input A	Input B	Σ	C$_{out}$	Carry In	Input A	Input B	Σ	C$_{out}$
0	0	0			1	0	0		
0	0	1			1	0	1		
0	1	0			1	1	0		
0	1	1			1	1	1		

6.6   Implementation 2

1) Open the Multisim file *Digital_Exp_06_Part_01b*.

2) Start the simulation.

3) Use the "A", "B", and "C" keys to open and close the "Input A", "Input B", and "Carry In" switches for each of the settings shown in Table 6-3. Alternatively, you can click on the switches to open or close them.

4) Record the state of the "Sum" and "Carry Out" probes for each combination of switch settings in the "Σ" and "C$_{out}$" columns of Table 6-3.

5) Stop the simulation.

Table 6-3: Observed Outputs for Full Adder Implementation 2

Carry In	Input A	Input B	Σ	C$_{out}$	Carry In	Input A	Input B	Σ	C$_{out}$
0	0	0			1	0	0		
0	0	1			1	0	1		
0	1	0			1	1	0		
0	1	1			1	1	1		

Conclusions for Part 1

Questions for Part 1

1) How could you configure a full adder to operate as a half-adder?

2) Do the outputs of the full adders in Table 6-2 and Table 6-3 match the expected values in Table 6-1?

3) What advantage, if any, does the full adder in *Digital_Exp_06_Part_01b* have over the full adder in *Digital_Exp_06_Part_01a*?

Part 2:   Parallel Binary Adders

The 7483 is a 4-bit parallel binary adder that can add two sets of 4-bit binary numbers. This device has carry in (C0) and carry out (C4) pins that allow you to cascade two or more devices to form 8-bit and wider adders.

6.7   Cascaded 8-Bit Adder

1) Calculate the sum for the "Input A" and "Input B" values for each row in Table 6-4 and record your answer as a three-digit hexadecimal value in the "Calculated Total" column. The first value is already given.

2) Open the Multisim file *Digital_Exp_06_Part_02a*.

3) Start the simulation.

4) Use the "Input A" and "Input B" DIP switches to apply each of the hexadecimal inputs in Table 6-4 to the adder circuit.

5) For each setting in Table 6-4, record the "Overflow" state as "ON" or "OFF" and the hexadecimal "Sum" value. What is the largest value the circuit can display without an overflow indication?

6) Stop the simulation.

Table 6-4: Results for 8-bit Adder (Values are in hexadecimal)

Input A	Input B	Calculated Total	Overflow	Sum	Input A	Input B	Calculated Total	Overflow	Sum
00	00	000			80	80			
1F	2B				AA	BB			
39	64				CA	FE			
7F	80				FF	FF			

Compare your "Calculated Total" value with the "Overflow" and "Sum" values for each row in Table 6-4. What does the binary value of "Overflow" indicator represent?

6.8   4-Bit Binary-to-BCD Converter

1) For each binary value in Table 6-5, indicate the corresponding 2-digit BCD value in the "BCD Value" column. The first value is already given.

2) Open the Multisim file *Digital_Exp_06_Part_02b*. Note that three of the BCD value probes connect to ground and never light up. Why are these probes included in the circuit if they are always OFF?

3) Start the simulation.

4) Use the DIP switch to apply each of the binary values in Table 6-5 to the converter circuit and record the resulting BCD pattern in the "Observed BCD" column.

5) Stop the simulation.

Table 6-5: Binary and BCD Equivalents

Binary	BCD Value	Observed BCD	Binary	BCD Value	Observed BCD
0000	0000 0000		1000		
0001			1001		
0010			1010		
0011			1011		
0100			1100		
0101			1101		
0110			1110		
0111			1111		

6.9   8-bit Binary-to-Decimal Converter

The 4-bit binary-to-BCD converter can be expanded to an 8-bit binary-to-decimal converter, although the design approach is more complex. In effect, the circuit independently adds the ones, tens, and hundreds represented by each bit in binary. It uses the basic circuitry for the 4-bit binary-to-BCD converter to adjust each sum as needed to keep the result in BCD, and adds the appropriate carries for each BCD digit to the next higher digit. Each BCD value is then displayed as a decimal digit.

1) Open the Multisim circuit *Digital_Exp_06_Part_02c*.

2) Start the simulation.

3) Use the DIP switch to apply each of the binary values in Table 6-6 to the converter circuit and record the resulting decimal value in the "Observed Value" column.

4) Stop the simulation.

Table 6-6: Corresponding Binary and Decimal Values

Binary Value	Decimal Value	Observed Value	Binary Value	Decimal Value	Observed Value
00000001			10101010		
00010000			11000011		
01100100			11111111		

How do the conversion times for the combinational binary-to-decimal converter compare with the conversion time for the sequential binary-to-decimal converter in Experiment 2?

Conclusions for Part 2

Questions for Part 2

1) The AND gates in *Digital_Exp_06_Part_02a* serve to buffer the DIP switch settings for the hexadecimal displays. One input of each 2-input AND gate between the input DIP switches and rest of the circuit is tied to V_{CC} (logic 1). Does this change the logic state of the DIP switch settings seen by the adders?

2) How would you modify the circuit in *Digital_Exp_06_Part_02a* to create a 12-bit adder? What additional parts would the circuit require?

3) How does the converter circuit of *Digital_Exp_06_Part_02b* convert the 4-bit binary value from the DIP switch to a BCD value?

4) Why do the digits on the displays in *Digital_Exp_06_Part_02c* sometimes briefly "flicker" when you change the DIP switch settings from one binary number to another?

Digital Experiment 7 - Encoders, Decoders, and Parity Checkers

7.1 Introduction

The full adder was one type of logic function. Other logic functions include encoding, decoding, and parity checking. Encoders use combinational logic to output a specific bit pattern to represent input information. Decoders use combinational logic to determine the information that a specific input bit pattern represents. Parity checkers provide a means of verifying that the representation of data is correct by generating a parity bit that depends on the data bits and then verifying that the parity bit and data bits are still consistent when the data is accessed. Parity can be even or odd, depending on whether the total number of 1s in the parity and data bits is an even or odd number.

In Part 1 of this experiment, you will observe the operation of encoders and decoders and how together they can convert information from one form to another. In Part 2, you will examine the operation of a parity checking circuit.

7.2 Reading

Floyd, *Digital Fundamentals, 10th Edition*, Chapter 6

7.3 Key Objectives

Part 1: Use an encoder to convert an input into a binary value, use a decoder to convert a binary pattern into a numeric pattern, and combine an encoder and decoder to convert an input into a numeric pattern.

Part 2: Examine the operation of the word generator and use it to verify the operation of a parity checking circuit.

7.4 Multisim Files

Part 1: *Digital_Exp_07_Part_01a*, *Digital_Exp_07_Part_01b*, and *Digital_Exp_07_Part_01c*

Part 2: *Digital_Exp_07_Part_02*

Part 1: Encoders and Decoders

7.5 Encoder Operation

1) Table 7-1 shows the inputs and outputs required for a 4-line-to-2-bit encoder. The encoder outputs a unique 2-bit binary pattern, represented by B_0 and B_1, that indicates which of four input switches, labeled SW_0 through SW_3, is open (indicated by a "1" in the SW_1 through SW_3 columns). Use the information to determine the encoding logic for output bits B_0 and B_1 as follows:

  a. If output bit B_1 for a row is a 1, record in the "B_1 Open Switch" column which switch is open for that row. If output bit B_1 for a row is 0, record a "0".

  b. If output bit B_0 for a row is a 1, record in the "B_0 Open Switch" column which switch is open for that row. If output bit B_0 for a row is 0, record a "0".

  c. OR each non-zero term in the "B_1 Open Switch" column to determine the encoding logic for B_1. Record this OR expressions next to "$B_1 =$" in the row labeled "Encoding Logic".

  d. OR each non-zero term in the "B_0 Open Switch" column to determine the encoding logic for B_0. Record this OR expressions next to "$B_0 =$" in the row labeled "Encoding Logic".

The first two rows of Table 7-1 are done for you as an example.

Table 7-1: 4-Line-to-2-Bit Encoder Logic Requirements

SW_0	SW_1	SW_2	SW_3	B_1	B_1 Open Switch	B_0	B_0 Open Switch
1	0	0	0	0	0	0	0
0	1	0	0	0	0	1	SW_1
0	0	1	0	1		0	
0	0	0	1	1		1	
Encoding Logic				$B_1 =$		$B_0 =$	

2) Open the Multisim file *Digital_Exp_07_Part_01a*. Does the encoding logic for the B_0 and B_1 probes match your expressions for B_0 and B_1 in the "Encoding Logic" row of Table 7-1?

3) Start the simulation.

4) Use the "0" through "3" keys to open and close the switches so that they correspond to the settings in Table 7-2. Record the states of the B_0 and B_1 probes. Do the observed states for B_0 and B_1 in Table 7-2 match the encoder requirements for B_0 and B_1 in Table 7-1?

5) Stop the simulation.

Table 7-2: Observed Encoder Output States

Switch Setting (Closed = 0, Open = 1)				Probe States (OFF = 0, ON = 1)	
SW_0	SW_1	SW_2	SW_3	B_0	B_1
0	0	0	0		
1	0	0	0		
0	1	0	0		
0	0	1	0		
0	0	0	1		

7.6 Decoder Operation

1) Table 7-3 shows the inputs for a 2-bit to 7-segment decoder. For the indicated combinations of input bits B_0 and B_1, the decoder outputs will activate the segments shown in Figure 7-1 to display a decimal digit from "0" to "3" on a standard 7-segment digital display. For each row in Table 7-3, record a "1" in the column for each segment that must be ON and a "0" in the column for each segment that must be OFF to display the required digit, as shown in Figure 7-1.

Table 7-3: 2-Bit-to-7-Segment Decoder Requirements

B_1	B_0	Digit	A	B	C	D	E	F	G
0	0	0							
0	1	1							
1	0	2							
1	1	3							

Figure 7-1: 7-Segment Patterns for 0 through 3

The decoder for this experiment will actually consist of an input decoder and output encoder. The input decoder will decode the binary inputs B_0 and B_1 to determine the represented digit. The output encoder will then encode the represented digit to determine the active display segments.

2) Use the data from Table 7-3 to complete each column of Table 7-4 as follows:

    a.   If B_1 or B_0 is a "1" for the indicated digit, write the true (noninverted) form of that bit in the column for that digit. If B_1 or B_0 is a "0", write the complemented (inverted) form of the input bit.

    b.   AND the terms in each column to determine the expression for the decoding logic of each digit and record this in the row labeled "Decoding Logic".

The first column is done for you as an example.

Table 7-4: Binary Input Decoder Logic

Digit	0	1	2	3
B_1	B_1'			
B_0	B_0'			
Decoding Logic	$B_1'B_0'$			

3) Use the data from Table 7-4 to complete Table 7-5 as follows:

    a.   For each row, write the digit (0, 1, 2, or 3) in a segment column if the segment is ON for that digit. If the segment is OFF for that digit, leave the cell blank.

    b.   Determine the decoding logic for each segment by ORing the terms in each column and record this in the row labeled "Encoding Logic".

The first column is done for you as an example.

Table 7-5: 7-Segment Output Encoder Logic

Digit	A	B	C	D	E	F	G
0	0						
1							
2	2						
3	3						
Encoding Logic	0+2+3						

4) Open the Multisim file *Digital_Exp_07_Part_01b*. Does the U_1 and U_2 input decoding logic match your expressions for each digit in "Decoding Logic" row of Table 7-4?

5) Verify that the encoding logic for each segment matches your expression for that segment in the "Encoding Logic" row of Table 7-5.

6) Start the simulation.

7) Use the "0" and "1" keys to open and close the switches so that they correspond to the settings in Table 7-6. Record the states of the "A" through "G" segments. Do the states of the "A" through "G" segments in Table 7-6 match the required states of the "A" through "G" segments in Table 7-3?

8) Stop the simulation.

Table 7-6: Observed 2-Bit to 7-Segment Decoder Output States (OFF = 0, ON = 1)

B1	B0	A	B	C	D	E	F	G
0	0							
0	1							
1	0							
1	1							

7.7   Combined Encoder and Decoder

1) Open the Multisim file *Digital_Exp_07_Part_01c*. This circuit connects the output of the encoder circuit in Section 7.5 to the input of the decoder circuit in Section 7.6.

2) Use the "0" through "3" keys to open and close the switches so that they correspond to the settings in Table 7-7. Record the states of the "A" through "G" segment probes. Do the states of the segment probes correspond to the digits represented by the open switches?

Table 7-7: Combined Encoder and Decoder Observed Outputs

SW$_0$	SW$_1$	SW$_2$	SW$_3$	A	B	C	D	E	F	G
0	0	0	0							
1	0	0	0							
0	1	0	0							
0	0	1	0							
0	0	0	1							

Conclusions for Part 1

Questions for Part 1

1) For the encoder circuit in *Digital_Exp_07_Part_01a*, why is the output of SW_0 connected only to the "0" logic probe and not to anything else?

2) For the decoder circuit in *Digital_Exp_07_Part_01b*, why is the logic probe for the "B" segment connected directly to V_{CC}?

3) For the combined encoder and decoder circuit in *Digital_Exp_07_Part_01c*, why are the segment states the same when all the switches are closed and when SW_0 is open?

Part 2: The Parity Checker

1) For each of the 4-bit data values D_0 through D_3 in Table 7-8, determine the parity bit value that will make the total of number of 1s an even value. Record your answer in the "Predicted" column of Table 7-8.

2) Open the Multisim file *Digital_Exp_07_Part_02*.

3) Start the simulation.

4) Use the DIP switch to apply each of the data bit combinations in Table 7-8 to the parity checker circuit. Record the state of the "Parity" logic probe in the "Observed" column of Table 7-8. How do the "Predicted" and "Observed" values compare?

Table 7-8: Predicted and Observed Even Parity Values

D_3	D_2	D_1	D_0	Even Parity Value		D_3	D_2	D_1	D_0	Even Parity Value	
				Predicted	Observed					Predicted	Observed
0	0	0	0			1	0	0	0		
0	0	0	1			1	0	0	1		
0	0	1	0			1	0	1	0		
0	0	1	1			1	0	1	1		
0	1	0	0			1	1	0	0		
0	1	0	1			1	1	0	1		
0	1	1	0			1	1	1	0		
0	1	1	1			1	1	1	1		

5) Press "0" to open the switch on the D_0 line (or, alternatively, click on the switch) to simulate a fault in the data transmission line.

6) Use the DIP switch to apply each of the data bit combinations in Table 7-9 to the parity checker circuit and record the state of the data bit and "Error" logic probes in Table 7-9. Does the parity checker detect every error?

Table 7-9: Observed Parity and Error Values for Single Fault Conditions

Transmit Switch Values				Receive Probe Values				Single Fault	
D_3	D_2	D_1	D_0	D_3	D_2	D_1	D_0	Parity	Error
0	0	0	0						
0	0	0	1						
0	0	1	0						
0	0	1	1						
0	1	0	0						
0	1	0	1						
0	1	1	0						
0	1	1	1						
1	0	0	0						
1	0	0	1						
1	0	1	0						
1	0	1	1						
1	1	0	0						
1	1	0	1						
1	1	1	0						
1	1	1	1						

7) Press "1" to open the switch on the D_1 line (or, alternatively, click on the switch) to simulate a double fault in the transmission lines.

8) Use the DIP switch to apply each of the data bit combinations in Table 7-10 to the parity checker circuit. Does the parity checker detect every error?

Table 7-10: Observed Parity and Error Values for Double Fault Conditions (continued on next page)

Transmit Switch Values				Receive Probe Values				Double Fault	
D_3	D_2	D_1	D_0	D_3	D_2	D_1	D_0	Parity	Error
0	0	0	0						
0	0	0	1						
0	0	1	0						
0	0	1	1						
0	1	0	0						

Transmit Switch Values				Receive Probe Values				Double Fault	
D_3	D_2	D_1	D_0	D_3	D_2	D_1	D_0	Parity	Error
0	1	0	1						
0	1	1	0						
0	1	1	1						
1	0	0	0						
1	0	0	1						
1	0	1	0						
1	0	1	1						
1	1	0	0						
1	1	0	1						
1	1	1	0						
1	1	1	1						

9) Open the switches on all the data lines and the parity line. Does the circuit ever indicate an error?

10) Stop the simulation.

Conclusions for Part 2

Questions for Part 2

1) How could you modify the circuit of *Digital_Exp_07_Part_02* to check for odd parity?

2) Based on your findings in Table 7-9 and Table 7-10, what is the limitation of the parity checker?

Digital Experiment 8 - Latches and Flip-Flops

8.1   Introduction

Most of the combinational logic circuits in previous experiments have been circuits for which the inputs directly affect the output. If applying an input changed the state of the output, then removing it would cause the output to revert to its previous state. This experiment introduces two types of latching devices. A latching device is one for which the output can remain in a specific state even after the input that produced the change is removed. Because latching devices effectively store information, they are basic memory devices.

The first type of device is the latch. There are several different types of latches, but all use cross-coupled feedback. This feedback connects the output of the device to its input, so that the same logic level as that which caused the output change is applied to the input. If the original input is removed, the feedback keeps the output in its latched state. A variation of the basic latch includes the gated latch, which requires an additional enable input to be HIGH or LOW to allow the control inputs to affect the latch. The addition of an edge-detection circuit to the enable input changes the latch into an edge-triggered device, for which a rising or falling edge applies the control inputs to the device.

The second type of device is the flip-flop. The flip-flop is similar to the latch, but differs in that it uses a clock rather than an enable input and its operation is more dynamic, emphasizing changes on the outputs.

In Part 1 of this experiment, you will investigate and compare the operation of several types of latches and an edge-triggered device. In Part 2, you will learn to operate the word generator and use it to verify the operation of a J-K flip-flop.

8.2   Reading

Floyd, *Digital Fundamentals, 10th Edition*, Chapter 7

8.3   Key Objectives

Part 1:   Compare and contrast the operation of basic and enabled S-R and S'-R' latches and verify the operation of an edge-triggered device.

Part 2:   Demonstrate the operation of the Multisim word generator and use it to observe operation of the J-K flip-flop.

8.4   Multisim Files

Part 1:   *Digital_Exp_08_Part_01a* through *Digital_Exp_08_Part_01f*

Part 2:   *Digital_Exp_08_Part_02a* and *Digital_Exp_08_Part_02b*

Part 1:   Basic Latch Operation

8.5   The S-R Latch

The S-R latch is the basis of all other latches. It consists of two NOR gates using cross-coupled feedback to latch information. The S-R latch has two active-HIGH inputs, called the *S* (for *set*) and *R* (for *reset*). It also has two outputs, called the *Q* and *Q'* outputs, which are complements of each other so that if $Q = 0$, $Q' = 1$ and vice versa. When $S = 1$ and $R = 0$, they will set the latch so that $Q = 1$ and $Q' = 0$. When $S = 0$ and $R = 1$, they will reset the latch so that $Q = 0$ and $Q' = 1$. When $S = 0$ and $R = 0$, the latch remains in its current state.

1)   Open the Multisim file *Digital_Exp_08_Part_01a*.

2)   Start the simulation.

3)   What are the initial states of the *S*, *R*, *Q*, and *Q'* logic probes?

4)   Press the "S" key to close the "SET" switch. What are the states of the *S*, *R*, *Q*, and *Q'* logic probes?

5) Press the "S" key again to open the "SET" switch. What are the states of the S, R, Q, and Q' logic probes?

6) Press the "R" key to close the "RESET" switch. What are the states of the S, R, Q, and Q' logic probes?

7) Press the "R" key again to open the "RESET" switch. What are the states of the S, R, Q, and Q' logic probes?

8) Stop the simulation.

8.6 Invalid Inputs for the S-R Latch

Because the Q and Q' outputs of the S-R latch must be complements of each other, the S and R inputs must not both be active at the same time. In this section you will determine what happens if this restriction is violated.

1) Open the Multisim file *Digital_Exp_08_Part_01b*. This circuit is an S-R latch, but the S and R inputs are tied together so that when S = 0, R = 0 and when S = 1, R = 1.

2) Start the simulation.

3) What are the initial states of the S, R, Q, and Q' logic probes?

4) Press the space bar to close the "INPUT" switch. What are the states of the S, R, Q, and Q' logic probes?

5) Press the space bar again to open the "INPUT" switch. What are the states of the S, R, Q, and Q' logic probes?

6) Stop the simulation.

8.7 The S'-R' Latch

The S'-R' latch is similar to the S-R latch except that it consists of two NAND gates using cross-coupled feedback to latch information. Because the S'-R' latch uses NAND gates, it has two active-LOW inputs, called the S' (for active-LOW set) and R' (for active-LOW reset). When the S' = 0 and R' = 1, they will set the latch so that Q = 1 and Q' = 0. When S' = 1 and R' = 0, they will reset the latch so that Q = 0 and Q' = 1. When S' = 1 and R' = 1, the latch remains in its current state.

1) Open the Multisim file *Digital_Exp_08_Part_01c*.

2) Start the simulation.

3) What are the initial states of the S', R', Q, and Q' logic probes?

4) Press the "R" key to close the "RESET" switch. What are the states of the S', R', Q, and Q' logic probes?

5) Press the "R" key again to open the "RESET" switch. What are the states of the S', R', Q, and Q' logic probes?

6) Press the "S" key to close the "SET" switch. What are the states of the S', R', Q, and Q' logic probes?

7) Press the "S" key again to open the "SET" switch. What are the states of the S', R', Q, and Q' logic probes?

8) Stop the simulation.

8.8 Invalid Inputs for the S'-R' Latch

As with the S-R latch, the S' and R' inputs must not both be active at the same time. Because S' and R' are both active-LOW inputs, this means that they must not both be LOW at the same time. In this part of the experiment you will determine what happens if this restriction is violated.

1) Open the Multisim file *Digital_Exp_08_Part_01d*. This circuit is an S'-R' latch, but the S' and R' inputs are tied together so that when $S' = 0$, $R' = 0$ and when $S' = 1$, $R' = 1$.

2) Start the simulation.

3) What are the initial states of the S', R', Q, and Q' logic probes?

4) Press the space bar to close the "INPUT" switch. What are the states of the S', R', Q, and Q' logic probes?

5) Press the space bar again to open the "INPUT" switch. What are the states of the S', R', Q, and Q' logic probes?

6) Stop the simulation.

8.9 The Gated S-R Latch

The gated S-R latch adds another input to the basic S-R latch, called the *enable* input. The enable input must be at its active level, or asserted, for the S and R inputs to affect the state of the latch. When the enable input is at its inactive level, or negated, the S and R inputs are blocked so that they cannot affect the state of the latch. The ability of the enable input to "gate" the S and R inputs in this manner is the reason this latch is called a "gated S-R latch".

1) Open the Multisim file *Digital_Exp_08_Part_01e*. Which logic gates block and pass the S and R inputs through the circuit and to the latching gates?

2) Start the simulation.

3) What are the initial states of the S, R, Q, and Q' logic probes?

4) Press the "S" and "R" keys to open and close the "SET" and "RESET" switches. Describe what happens.

5) Use the "S" and "R" keys to open both the "SET and "RESET" switches.

6) Press the "E" key to close the "ENABLE" switch.

7) Use the "S" and "R" keys to open and close the "SET" and "RESET" switches. Describe the general operation of the circuit.

8) Stop the simulation.

8.10 Edge-Triggering

The gated S-R latch allows the *S* and *R* inputs to affect the latch as long as the enable input is active. A drawback of this is that the outputs may be unpredictable if the inputs are changing while the enable is active. Adding an edge-detection circuit can improve this by having the device latch the input data on the edge of the enable. The exclusive-OR gate can act as an edge-detector if the same signal is applied to both inputs, but the circuit slightly delays the signal to one input. For a short period of time the logic levels on the inputs are different, creating a short pulse on the output of the XOR gate that acts the same as the enable signal in a gated latch.

1) Open the Multisim file *Digital_Exp_08_Part_01f*.
2) Start the simulation.
3) What are the initial states of the *S*, *R*, *Q*, and *Q'* logic probes?

4) Use the "S" and "R" keys to open and close the "SET" and "RESET" switches and describe what happens.

5) Use the "S" and "R" keys to open both the "SET and "RESET" switches.
6) Press the "S" key to close the "SET" switch.
7) Press the "E" key to close the "ENABLE" switch. Describe what happens.

8) Press the "S" key to open the "SET" switch.
9) Press the "R" key to close and open the "RESET" switch. Describe what happens.

10) Press the "E" key to open the "ENABLE" switch. Describe what happens.

11) Press the "R" key to close the "RESET" switch.
12) Press the "E" key to close the "ENABLE" switch. Describe what happens.

13) Press the "R" key to open the "RESET" switch.
14) Press the "S" key to close and open the "SET" switch. Describe what happens.

15) Press the "E" key to open the "ENABLE" switch. Describe what happens.

16) Stop the simulation.

Conclusions for Part 1

Questions for Part 1

1) When $S = 0$, $R = 0$, $Q = 1$, and $Q' = 0$, how does the cross-coupled feedback keep the outputs from changing in the S-R latch circuit?

2) Why cannot both inputs of the S-R latch be active at the same time?

3) The S and R inputs of a gated S-R latch are both 0 when the enable input becomes active. The S input goes HIGH and then LOW while the enable input is active. Just before the enable input becomes inactive, the R input goes HIGH and then LOW. What are the final states of the Q and Q' outputs?

Part 2:   Flip-Flops

8.11   The Multisim Word Generator

The Multisim word generator does not exist as a real device, although its function can be duplicated with computers and software. The function of the word generator is almost the opposite that of the logic analyzer. The logic analyzer observes logic levels of multiple signals in a circuit and displays their values. With the word generator you specify the values of multiple signals and the word generator produces the signals. Figure 8-1 shows the word generator tool. Figure 8-2 shows the minimized and expanded views of the word generator.

Figure 8-1: The Word Generator Tool

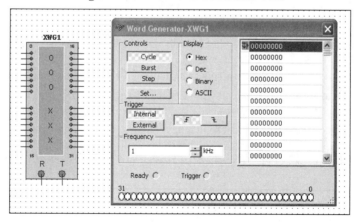

Figure 8-2: Word Generator Minimized and Expanded Views

To use the word generator, specify a sequence of HIGH (1) and LOW (0) logic levels for up to 32 channels, numbered from 0 to 31. To do so, click in the list of values in the pattern buffer window and enter the patterns you wish the generator to output. You can enter and display these values in hexadecimal, decimal, binary, or ASCII, depending upon the radio button you select in the **Display** section to the left of the window.

To the left of each value in the list is a box. If you right-click in the box next to a value, a menu will open. The menu allows you to specify the start (**Set Initial Position**) and end (**Set Final Position**) of the pattern you want the word generator to use, and the value at which you want execution to begin (**Set Cursor**). You can also set a breakpoint (**Set Break-Point**) that will pause the generator at that value so that you can examine the state of the circuit, or clear a previously set breakpoint (**Delete Break-Point**).

The **Controls** section allows you to specify how you want the word generator to operate. **Cycle** will repeatedly execute the pattern between the specified start and end values of the list until you stop the simulation or select **Burst** or **Step**. **Burst** will execute the values between the position of the cursor and the end value in the list and then pause execution. **Step** will execute one value at a time and then pause execution. Clicking on any of these buttons will automatically begin execution.

The **Set...** button permits you to clear the buffer, save the pattern that the buffer contains, load a saved pattern into the buffer, or load one of several pre-defined patterns into the buffer. It also allows you to specify the size of the buffer (up to 8192 values) and display the buffer size in decimal or hexadecimal.

The **Trigger** controls are similar to those of the logic analyzer and allow you to specify whether the rising or falling edge of an internal or external trigger is used to initiate execution.

The **Frequency** controls specify how rapidly the word generator outputs the values in the buffer. For example, a "1 kHz" setting will output one value in the buffer every millisecond.

The connections to the word generator consists of 32 data lines (lines 0 through 15 are on one side of the minimized instrument and lines 16 through 31 on the other side), a "Ready" output that indicates when the word generator is ready to begin execution, and a "Trigger" input that will initiate the execution when an external trigger is specified.

In this part of the experiment, you will program the word generator to output a 3-bit binary sequence from 000 to 111.

1) Open the Multisim file *Digital_Exp_08_Part_02a*.

2) Double-click on the word generator to expand it.

3) Click on the "Binary" radio button to display the contents of the pattern buffer window in binary.

4) Click on the first value in the buffer to select it and change the rightmost (least significant) 3 bits of the first value in the buffer to "000", if they are not already set to "000".

5) Repeat step 4 to change the rightmost (least significant) 3 bits of the next seven buffer values to "001", "010", "011", "100", "101", "110", and "111".

6) Right-click in the box to the left of the last entered value and select "Set Final Position".

7) Right-click in the box to the left of the first value and select "Set Cursor".

8) Double-click on the logic analyzer to expand it. If necessary, move the expanded views so that they do not block each other.

9) Click the "Burst" button in the **Controls** section of the word generator.

10) After the logic analyzer stops collecting data, click on the logic analyzer's "Clocks/Div" control to expand or reduce the display until you can see the entire collected waveform in one screen.

11) Verify that the collected data matches the binary patterns you entered into the word generator's pattern buffer. The display on the logic analyzer should be similar to that in Figure 8-3. If the patterns do not match, repeat steps 4 through 10.

12) Stop the simulation.

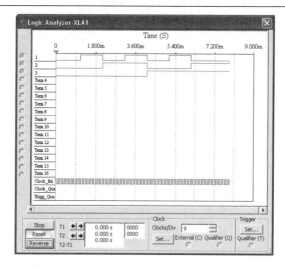

Figure 8-3: Collected Binary Waveform

8.12 The J-K Flip-Flop

In this section, you will observe the operation of a J-K flip-flop (a common device in sequential circuits) by using a timing diagram to program the word generator pattern buffer. Table 8-1 shows the next-state table (the sequential logic equivalent of the combinational logic truth table) for the J-K flip-flop. Note that an "X" (don't care") indicates that the state of the input does not matter and can be either a 0 or 1.

Table 8-1: J-K Flip-Flop Next-State Table (X = "Don't Care")

PR'	CLR'	J	K	CLK	Q_N	Q_{N+1}	Description
1	0	X	X	X	X	0	Asynchronous reset
0	1	X	X	X	X	1	Asynchronous set
1	1	0	0	↓	Q_N	Q_N	No change
1	1	0	1	↓	X	0	Synchronous clear
1	1	1	0	↓	X	1	Synchronous set
1	1	1	1	↓	Q_N	Q'_N	Synchronous toggle

As the Table 8-1 shows, the *PR*' (preset) and *CLR*' (clear) inputs are asynchronous inputs and will immediate set and clear the flip-flop, respectively, regardless of the state of the *CLK* (clock) input. The *J* and *K* inputs are synchronous inputs and require a falling edge on *CLK* to affect the output so that the output changes on the falling clock edge. Q_N is the state of the *Q* output just before the indicated inputs are applied, and Q_{N+1} is the state of the *Q* output just after the indicated inputs are applied. When *PR*' = 1, *CLR*' = 1, *J* = 0, and *K* = 0, for example, the *Q* output after the falling edge of the clock will be the same as it was before the falling edge of the clock. When *PR*' = 1, *CLR*' = 1, *J* = 1, and *K* = 0, the *Q* output will be 1 after the falling edge of the clock regardless of the state of the output before the falling edge of the clock.

1) For the timing diagram in Figure 8-4, write a "0" or "1" to indicate the state of the flip-flop inputs for each interval. The *CLK* input has been completed for you as an example.

2) Use the information in Table 8-1 to determine the state (0 or 1) of the *Q* output for each interval in Figure 8-4. Record your values.

3) Use your values for the *Q* output in Figure 8-4 to determine the state of the *Q*' output (the complement of the *Q* output). Record your values.

4) Sketch the timing diagram of the *Q* and *Q*' outputs for the J-K flip-flop in Figure 8-4 by drawing a line under every "0" and over every "1" and rising or falling edges to connect these levels.

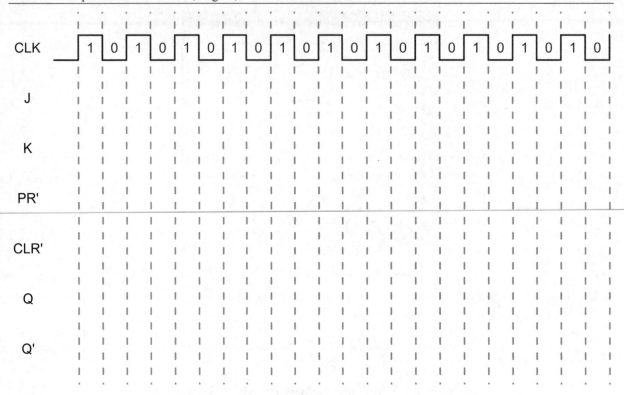

Figure 8-4: Timing Diagram for J-K Flip-Flop

5) Open the Multisim file *Digital_Exp_08_Part_02b*.

6) Double-click the word generator to expand it.

7) Select "Binary" in the **Display** section if it is not already selected.

8) Use your values in Figure 8-4 to input the binary pattern buffer data for the *CLK*, *J*, *K*, *PR'*, and *CLR'* inputs.

9) Double-click the logic analyzer to expand it. If necessary, move the expanded views so that they do not block each other.

10) Click the "Burst" button in the **Controls** section of the word generator.

11) After the logic analyzer stops collecting data, click on the logic analyzer's "Clocks/Div" controls to expand or reduce the display until you can see the entire collected waveform on one screen.

12) Compare the waveforms in the logic analyzer with those in Figure 8-4. If they do not match, repeat steps 8 through 10.

13) If you are unable to obtain the correct waveform, follow these steps:

    a.   Click the **Set...** button in the **Controls** section of the word generator.

    b.   Select the "Load" in the **Pre-set Patterns** section.

    c.   Click the **Accept** button to open the file browser.

    d.   The browser should open the folder that contains the *Digital_Exp_08_Part_02b* file. If not, navigate to the folder that contains it.

    e.   Select the *JK_Flop2.dp* data pattern file.

    f.   Click the **Open** button to load the pattern into the word generator buffer. Compare the values in the pattern buffer with your values in Figure 8-4.

    g.   Click the **Burst** button in the **Controls** section of the word generator. Compare the logic analyzer waveforms with your values in Figure 8-4.

14) Stop the simulation.

Conclusions for Part 2

Questions for Part 2

1) The pattern you entered into the word generator in Section 8.11 was a 3-bit binary up count, as the patterns were that of an increasing 3-bit binary count. If you wished to see a 4-bit binary down count, what pattern would you enter into the pattern buffer?

2) Suppose that when you applied the inputs in Figure 8-4 to a J-K flip-flop you found that the Q output was always LOW and that the Q' was always HIGH. What would you suspect is the most likely problem?

3) Suppose that when you applied the inputs in Figure 8-4 to a J-K flip-flop you found that the Q output was as expected, but that the Q' output was always 0. What possible problems would you suspect?

Name _____   Class _____

Date _____   Instructor_____

Digital Experiment 9 - Counters

9.1   Introduction

Counters belong to a class of digital circuits known as *state machines*. A state machine is a circuit that sequences through a series of states, or unique combination of inputs and outputs. A counter consists of a specific number of count states, which the counter represents with a defined pattern of 0s and 1s on its outputs. A 2-bit up counter, for example, shows the sequence 00, 01, 10, 11 on its outputs. A counter requires one flip-flop to implement each bit, and each bit has two states, so that N flip-flops can create an N-bit counter with a maximum of 2^N count states.

There are two major classes of counters: asynchronous counters and synchronous counters. In asynchronous counters, also called *ripple counters*, some of the flip-flops use the outputs of other flip-flops as their clock signal. Because a flip-flop exhibits a delay between its clock input and its outputs, the outputs of asynchronous counters do not change at the same time (which explains the name "asynchronous counter"). In the worst case scenario, there could be $N - 1$ propagation delays between the first output and last output as the output changes "ripple" through the counter. These delays can lead to transient count states as the counter outputs are settling to their final value for the count state and lead to a problem known as decoding glitches when the state is decoded. A logic analyzer is usually better than an oscilloscope at detecting these glitches, as logic analyzers typically have "glitch capture" circuitry that can recognize and display these very short pulses.

In synchronous counters, each flip-flop uses the same clock signal so that all the outputs change at virtually the same time. There is still a possibility of decoding glitches with synchronous counters because the clock-to-output delay for each flip-flop can differ, but the unwanted transient count states for synchronous counters are much shorter than those for asynchronous counters. If transient count states in a system are not acceptable, a Gray code counter in which only one output changes between adjacent count states is the best solution.

In Part 1 of this experiment, you will examine the structure and operation of basic asynchronous counters and investigate decoding glitches. In Part 2, you will investigate the operation of a counter application using a synchronous counter IC.

9.2   Reading

Floyd, *Digital Fundamentals, 10th Edition*, Chapter 8

9.3   Key Objectives

Part 1:   Analyze the operation of discrete asynchronous counters and the transient count states inherent in asynchronous counters.

Part 2:   Examine the operation of a typical IC synchronous counter application.

9.4   Multisim Files

Part 1:   *Digital_Exp_09_Part_01a*, *Digital_Exp_09_Part_01b*, and *Digital_Exp_09_Part_01c*

Part 2:   *Digital_Exp_09_Part_02*

Part 1:   Asynchronous Counters

9.5   3-Bit Up Counter

1)   Open the Multisim file *Digital_Exp_09_Part_01a*.

2)   What will happen when switch J_1 closes?

3) What will happen to each flip-flop when a clock edge occurs? Which clock edge (rising or falling) will affect the flip-flops?

4) Assume that the states of D_0, D_1, and D_2 are all 0, switch J_1 is open, and switch J_2 closes. Analyze the operation of the counter circuit as follows:

    a. Determine the effect of the clock on the next state D_0. When does D_0 change state?

    b. Determine the effect of D_0 changing on the next state of D_1. When does D_1 change state?

    c. Determine the effect of D_1 changing on the next state of D_2. When does D_2 change state?

5) Use your analysis in step 4 to complete Table 9-1. The first two rows are done for you as an example.

Table 9-1: Next-State Table for 3-Bit Up Counter

CLOCK	D_2		D_1		D_0	
	Present	Next	Present	Next	Present	Next
1	0	0	0	0	0	1
2	0	0	0	1	1	0
3						
4						
5						
6						
7						
8						

6) Start the simulation.

7) Close and open switch J_1 by pressing the "0" key or clicking on the switch. Record the initial states of the D_0, D_1, and D_2 logic probes in Table 9-2.

8) Close and open switch J_2 by pressing the "C" key or clicking on the switch. Record the new states of the D_0, D_1, and D_2 logic probes in Table 9-2.

9) Repeat Step 8 until you have completed Table 9-2.

10) Stop the simulation.

Table 9-2: 3-Bit Up Counter Logic Probe States (OFF = 0, ON = 1)

Switch Operation	D_2	D_1	D_0	Switch Operation	D_2	D_1	D_0
Initial State				5			
1				6			
2				7			
3				8			
4							

9.6 3-Bit Down Counter

1) Open the Multisim file *Digital_Exp_09_Part_01b*. How does this circuit differ from that of the 3-bit up counter?

2) Complete the "Predicted" columns of Table 9-3 by inverting the D_0 through D_2 values in Table 9-2.

3) Start the simulation.

4) Close and open switch J_1 by pressing the "0" key or clicking on the switch. Record the initial states of the D_0, D_1, and D_2 logic probes in the "Observed" columns of Table 9-3.

5) Close and open switch J_2 by pressing the "C" key or clicking on the switch. Record the new states of the D_0, D_1, and D_2 logic probes in the "Observed" Columns of Table 9-3.

6) Repeat Step 5 until you have completed Table 9-3.

7) Stop the simulation.

Table 9-3: 3-Bit Down Counter Logic Probe States (OFF = 0, ON = 1)

Switch Operation	Predicted			Observed		
	D_2	D_1	D_0	D_2	D_1	D_0
Initial State						
1						
2						
3						
4						
5						
6						
7						
8						

9.7 Decoding Glitches

1) Open the Multisim file *Digital_Exp_09_Part_01c*. This circuit adds a 3-to-8 line decoder to the 3-bit counter. Verify that the counter formed by the J-K flip-flops is identical to the Section 9.5 up counter.

2) Double-click on the logic analyzer to expand it.

3) Start the simulation. The logic analyzer will begin collecting data, but will not save the data until the input signals meet the trigger conditions.

4) Open switch J_1 by pressing the "0" key or clicking on the switch. After a short time the logic analyzer will detect the trigger and begin saving data. When the logic analyzer stops collecting data, stop the simulation.

5) Use the scroll bar beneath the logic analyzer display to scroll to the start of the data capture so that you can see the green line that indicates the detected trigger. Click the **Set...** button in the **Trigger** section of the logic analyzer controls to view the programmed trigger conditions. What trigger is specified in the **Trigger Patterns** section? Does the green line match the specified trigger conditions?

6) Use the ◄ and ► controls below the logic analyzer display to slowly scroll through the collected data. You will see some very short pulses, or glitches, on the decoder outputs. These glitches indicate that the decoder observed some very brief transient counter states on its inputs. Where do these glitches appear? Based on when these glitches occurred, explain what happened.

Conclusions for Part 1

Questions for Part 1

1) Do the D_0 through D_2 logic probes change states when switch J_1 opens or closes? Why?

2) Would you expect any decoding glitches to occur when the binary counter in *Digital_Exp_09_Part_01c* changed from an odd number to an even number? Why or why not?

Part 2:   A Counter Application

In this part of the experiment you will analyze the operation of a model rocket launching circuit that uses a 74LS190N synchronous decade counter IC. Refer to the description of the 74HC190 in the text or a datasheet for details on how the 74190 counter operates.

1) Open the Multisim file *Digital_Exp_09_Part_02*.

2) What is the purpose of switch J_2?

3) What happens when switch J_2 is open and switch J_3 is closed?

4) What will happen when switch J_2 is closed and J_3 is open?

5) Start the simulation.

6) Close and open switch J_3 by pressing the "L" key or clicking on the switch. What are the states of the logic probes?

7) Close switch J_2 by pressing the "A" key or clicking on the switch. What are the states of the logic probes as the countdown progresses?

Conclusions for Part 2

Questions for Part 2

1) How would you convert the circuit to one that would count up from 0000 to 1001?

2) What are two ways to abort a launch if the launch sequence is accidentally started?

3) Why does the circuit use NOR gates as inverters rather than actual inverters?

Digital Experiment 10 - Shift Registers

10.1 Introduction

An important class of sequential circuits based on the flip-flop is the shift register. As its name suggests, each flip-flop in a shift register can transfer, or "shift" its data to an adjacent flip-flop. Depending upon its design, a shift register can store, or load, data serially or in parallel, and output data serially or in parallel. Serial load and access operations work with one bit at a time and are therefore slower than parallel load and access operations that work with multiple bits at a time, but serial shift register designs and interface circuits are simpler. Flip-flops in a shift register are called *stages*.

Shift registers are used in a variety of applications. Binary multipliers and dividers use shifters to multiply and divide binary values by powers of two, and shift registers are used for serial-to-parallel and parallel-to-serial conversions. Special counters based on shift registers, called *ring* and *Johnson* counters, are used to generate test patterns to verify the operation of digital circuits or enable specific parts of digital systems at the right time.

In Part 1 of this experiment, you will examine the operation of serial in, parallel out (SIPO) shift registers. In Part 2, you will investigate the characteristics of ring and Johnson counters.

10.2 Reading

Floyd, *Digital Fundamentals, 10th Edition*, Chapter 9

10.3 Key Objectives

Part 1:   Examine the basic load, shift, and output characteristics of unidirectional and bidirectional SIPO shift registers.

Part 2:   Investigate and compare the operation of ring and Johnson counters.

10.4 Multisim Circuits

Part 1:   *Digital_Exp_10_Part_01a* and *Digital_Exp_10_Part_01b*

Part 2:   *Digital_Exp_10_Part_02*

Part 1:   Basic Shift Register Operation

10.5 The Unidirectional SIPO Shift Register

The unidirectional SIPO shift register loads data serially and allows stored data to be accessed in parallel, so that data is stored one bit at a time and output simultaneously. A unidirectional shift register can shift data in only one direction, so that data always moves through the internal flip-flops in the same order.

1) Open the Multisim file *Digital_Exp_10_Part_01a*.

2) Start the simulation.

3) Open and close switch J_3 by pressing the "0" key or clicking on the switch. Record the initial states of the logic probes in the State 0 row of Table 10-1.

4) For the State 1 through 14 rows in Table 10-1

   a.   Press the "D" key or click on the switch J_1 to set the data switch to the indicated position (OPEN or CLOSED).

   b.   Open and close J_2 to clock the shift register.

   c.   Record the state of the logic probes after clocking the shift register.

5) Open and close switch J_3 by pressing the "0" key or clicking on the switch. Record the final states of the logic probes in the State 15 row of Table 10-1.

6) Stop the simulation.

Table 10-1: Unidirectional Shift Register Outputs (OFF = 0, ON = 1)

State	J_1	X_1	X_2	X_3	X_4	State	J_1	X_1	X_2	X_3	X_4
0	Initial					8	CLOSED				
1	OPEN					9	OPEN				
2	OPEN					10	CLOSED				
3	OPEN					11	CLOSED				
4	OPEN					12	OPEN				
5	CLOSED					13	OPEN				
6	CLOSED					14	CLOSED				
7	OPEN					15	Final				

10.6   The Bidirectional SIPO Shift Register

The bidirectional SIPO shift register is similar to the unidirectional SIPO shift register, but it adds an additional control for the direction in which data shifts through the register.

1) Open the Multisim file *Digital_Exp_10_Part_01b*.

2) Start the simulation.

3) Open and close switch J_3 by pressing the "0" key or clicking on the switch. Record the initial states of the logic probes in the State 0 row of Table 10-2.

4) For the State 1 through 14 rows in Table 10-2

   a. Press the "D" key or click on the switch J_1 to set the data switch to the indicated position (OPEN or CLOSED).

   b. Press the "R" key or click on switch J_4 to set the direction switch to the indicated position (OPEN or CLOSED).

   c. Open and close J_2 to clock the shift register.

   d. Record the state of the logic probes after clocking the shift register.

5) Open and close switch J_3 by pressing the "0" key or clicking on the switch. Record the final states of the logic probes in the State 15 row of Table 10-2.

6) Stop the simulation.

Table 10-2: Bidirectional Shift Register Outputs (OFF = 0, ON = 1) (continued on next page)

State	J_1	J_4	X_1	X_2	X_3	X_4
0	Initial					
1	OPEN	CLOSED				
2	CLOSED	CLOSED				
3	OPEN	CLOSED				
4	CLOSED	OPEN				
5	CLOSED	OPEN				
6	OPEN	OPEN				
7	OPEN	CLOSED				
8	CLOSED	CLOSED				

State	J_1	J_4	X_1	X_2	X_3	X_4
9	OPEN	CLOSED				
10	CLOSED	CLOSED				
11	CLOSED	CLOSED				
12	OPEN	OPEN				
13	OPEN	OPEN				
14	CLOSED	OPEN				
15	Final					

Conclusions for Part 1

Questions for Part 1

1) Is the "CLEAR" control for the unidirectional shift register, corresponding to switch J_3, synchronous or asynchronous? Explain your answer.

2) From the data in Table 10-2, in which direction does data shift when switch J_4 is OPEN? What logic level does this represent?

3) What is the source of data for flip-flop U_{4A} when data shifts to the right? What is the source of data for flip-flop U_{4A} when data shifts to the left?

Part 2:   Shift Register Applications

Ring and Johnson counters allow the contents of shift registers to continuously rotate through the registers by connecting the output of the last stage to the input of the first stage. The difference between the ring counter and the Johnson counter is that the last stage in a Johnson counter inverts the output before applying it to the first stage. This is why a Johnson counter is also called a "twisted ring" counter.

10.7 Ring and Johnson Counters

1) Open the Multisim file *Digital_Exp_10_Part_02*.
2) Start the simulation.
3) Close and open switch J_3 by pressing the "0" key or clicking on the switch.
4) Set input 1 of switch J_1 to "1" (to the left) and inputs 2 through 8 to "0" (to the right) to create the pattern "10000000".
5) Close switch J_4 to connect the clock to the shift registers by pressing the "C" key or clicking the switch.
6) Close switch J_2 by pressing the "L" key or clicking the switch to load the shift registers. What pattern appears on logic probes X_1 through X_8?

7) Open switch J_2 by pressing the "L" key or clicking the switch to enable the shift operation of the shift registers. With switch J_4 closed and the other switches open, what sequence of patterns appears on logic probes X_1 through X_8? How many unique patterns appear?

8) Close switch J_2 by pressing the "L" key or clicking the switch to reload the shift registers.

9) Close switch J_5 by pressing the "J" key or clicking the switch to change the mode of the counter.

10) Open switch J_2 by pressing the "L" key and clicking the switch to enable the shift operation of the shift registers. What sequence of patterns now appears on logic probes X_1 through X_8? How many unique patterns appear?

11) Stop the simulation.

Conclusions for Part 2

Questions for Part 2

1) How does the setting of switch J_5 affect the operation of the counter?

2) What is the advantage of the Johnson counter over a standard ring counter?

3) Suppose when you simulate the circuit in *Digital_Exp_10_Part_02* you find that the circuit always operates as a Johnson counter. What fault would you suspect?

Digital Experiment 11 - Memory Devices and Operation

11.1 Introduction

In digital electronics, a memory device is any device that can store binary data. Latches, flip-flops, and registers are examples of basic memory devices that you investigated in previous experiments. However, although they can store data, their direct application as practical memory devices is very limited. To be practical, memory devices should 1) store information indefinitely, 2) internally organize data to reliably store and/or retrieve specific data, and 3) support memory expansion so that adding more devices can increase the total amount of stored data.

There are many varieties of memory, but the two basic types are memory from which data is only retrieved during normal operation, and memory that both stores and retrieves data during normal operation. Memory from which data normally is retrieved, or read, is called read-only memory (ROM). Memory that allows users to both store (write) and retrieve (read) data is called random access memory (RAM) or, less commonly, read/write memory (R/WM). Static RAM (SRAM) stores each data bit in a latch, and will retain information as long as power is applied. Dynamic RAM (DRAM) stores each data bit as charge on a capacitor and can store information only as long as the charge is periodically refreshed before it leaks off the capacitor. SRAM is typically faster and used in cache and other high-speed memory applications, but DRAM has a lower cost per bit and more widely used for general system memory.

In Part 1 of this experiment, you will examine memory operations by manually applying the standard signals that address, read, and write data to a memory circuit. In Part 2, you will examine how memory circuits are expanded either to increase the data width or the number of memory locations of basic memory devices.

11.2 Reading

Floyd, *Digital Fundamentals, 10th Edition*, Chapter 10

11.3 Key Objectives

Part 1:   Investigate how data, address, read, and write signals store and access information in a memory device.

Part 2:   Compare and contrast the expansion of memory by increasing the data width and increasing the number of memory locations.

11.4 Multisim Circuits

Part 1:   *Digital_Exp_11_Part_01*

Part 2:   *Digital_Exp_11_Part_02a* and *Digital_Exp_11_Part_02b*

Part 1:   Basic Memory Operations

1)   Open the Multisim file *Digital_Exp_11_Part_01*.

2)   Start the simulation.

3)   Attempt to write the pattern "01" into address location 0 and "10" into address location 1 as follows:

    a.   Use the "0" key (or left-click with the mouse) to open the "D0" switch.

    b.   Use the "1" key to close the "D1" switch.

    c.   Use the "A" key to close the "A0" switch.

    d.   Use the "W" key to close and open the "WRITE'" switch.

    e.   Close the "D0" switch.

    f.   Open the "D1" switch.

    g.   Open the "A0" switch.

    h.   Close and open the "WRITE'" switch.

4)   Attempt to read back the data from memory by repeating step 3, but use the "R" key rather than "W" key to close and open the "READ'" switch rather than the "WRITE'" switch. What happens?

5)   Attempt to write the pattern "01" in address location 0 and "10" in address location 1 as follows:
   a.   Open the "D0" switch.
   b.   Close the "D1" switch.
   c.   Close the "A0" switch.
   d.   Use the "S" key to close the "SELECT'" switch.
   e.   Close and open the "WRITE'" switch.
   f.   Open the "SELECT'" switch.
   g.   Close the "D0" switch.
   h.   Open the "D1" switch.
   i.   Close the "A0" switch.
   j.   Close the "SELECT'" switch.
   k.   Close and open the "WRITE'" switch.
   l.   Open the "SELECT'" switch.

6)   Attempt to read back the data from memory by repeating step 5, but use the "R" key rather than the "W" key to close and open the "READ'" switch rather than the "WRITE'" switch. What happens?

7)   Attempt to read back the data from memory by repeating Step 5, but do not close the "SELECT'" switch. What happens?

8)   Attempt to write the pattern "10" in address location 0 and "01" in address location 2 as follows:
   a.   Close the "D0" switch.
   b.   Open the "D1" switch.
   c.   Close the "A0" switch.
   d.   Close the "WRITE'" switch.
   e.   Close and open the "SELECT'" switch.
   f.   Open the "WRITE'" switch.
   g.   Open the "D0" switch.
   h.   Close the "D1" switch.
   i.   Open the "A0" switch.
   j.   Close the "WRITE'" switch.
   k.   Close and open the "SELECT'" switch.
   l.   Open the "WRITE'" switch.

9)   Attempt to read back the data from memory by repeating step 8, but use the "R" key rather than "W" key to close and open the "READ'" switch rather than the "WRITE'" switch. What happens?

10) Stop the simulation.

Conclusions for Part 1

Questions for Part 1

1) What is the function of the SELECT' control line for the memory circuit?

2) For the memory circuit in *Digital_Exp_11_Part_01*, what are two ways to write data into a memory location?

3) From the design of the circuit in *Digital_Exp_11_Part_01*, what would happen if the address line A0 changed before the WRITE' and SELECT' control lines went HIGH?

Part 2:   Memory Expansion

11.5 Expanding the Data Width

1) Open the Multisim file *Digital_Exp_11_Part_02a*. What are the similarities and differences of this circuit compared to that of *Digital_Exp_11_Part_01*?

2) Start the simulation.
3) Use the same read and write processes as for the circuit in *Digital_Exp_11_Part_01* to apply data, address, and control signals to the circuit. What happens?

4) Stop the simulation.

11.6 Increasing the Number of Memory Locations

1) Open the Multisim file *Digital_Exp_11_Part_02b*. What are the similarities and differences of this circuit compared to that of *Digital_Exp_11_Part_01*?

2) Start the simulation.

3) Use the same read and write processes as for the circuit in *Digital_Exp_11_Part_01* to apply data, address, and control signals to the circuit. What happens?

4) Stop the simulation.

Conclusions for Part 2

Questions for Part 2

1) How do the basic memory operations change when memory is expanded?

2) How many latches would be necessary to expand the circuit to support four data bits and two address lines?

3) Each of the latch outputs has a tri-state buffer that places its output into a high-impedance state (similar to an open switch to its internal circuitry and the external pin) so that it does not output a voltage. Why is this necessary?

Digital Experiment 12 - Programmable Logic Concepts

12.1 Introduction

Fixed-function logic is logic whose function cannot change. An AND gate will always function as an AND gate so that its output is HIGH when all its inputs are HIGH and LOW if any input is LOW. If fixed-function logic devices do not provide the function a circuit requires, a designer must externally connect them to do so. The user can configure programmable logic to provide the desired function.

Programmable logic architectures are divided into two main groups: programmable logic devices (PLDs) and field-programmable gate arrays (FPGAs). PLD architectures consist of an array of AND gates, an array of OR gates, and configurable output macrocells, and allow the user to selectively connect inputs to the AND gates, selectively connect the AND outputs to the OR gates, or both. FPGA architecture consists of configurable logic blocks, configurable input/output blocks, and programmable interconnects. Signal paths in PLDs typically consist of an input, an AND gate, and an OR gate, so that propagation delays are much more consistent and predictable. FPGAs provide much greater density and complexity, but signal timing is much less predictable and requires extensive simulation and testing to verify that designs will not have race conditions or other timing-related problems.

In this experiment you will use switches to simulate making the internal circuit connections that programmable logic uses to configure the internal logic of a device. In Part 1 of this experiment, you will simulate programming a PLD macrocell to implement specific logic functions. In Part 2, you will simulate programming a ROM look-up table (LUT), similar to those in some FPGAs, to implement sum-of-products (SOP) logic functions.

12.2 Reading

Floyd, *Digital Fundamentals, 10th Edition*, Chapter 11.

12.3 Key Objectives

Part 1:   Demonstrate how a PLD macrocell is programmed to implement desired logic functions.

Part 2:   Demonstrate how a ROM look-up table is used to implement a desired SOP expression.

12.4 Multisim Files

Part 1:   *Digital_Exp_12_Part_01*

Part 2:   *Digital_Exp_12_Part_02*

Part 1:   Example PLD Macrocell

12.5 Programming a 2-Input Exclusive-OR Function

1)   Open the Multisim file *Digital_Exp_12_Part_01*. The circuit simulates the macrocell of a 3-input PLD.

- The three SPST switches J_1, J_2, and J_3 apply HIGH and LOW levels for inputs A, B, and C to the macrocell. A CLOSED switch is a LOW (0) and an OPEN switch is a HIGH (1).

- The six switch packages J_4 through J_9 connect the six possible input terms (A, A', B, B', C, and C') to the four AND gates, which are connected in turn to a 4-input OR gate. The first (leftmost) switch of each package connects the input to AND gate U_1, the second to AND gate U_2, the third to AND gate U_3, and the fourth (rightmost) to AND gate U_4. If the corresponding switches of the switch packages are both in the UP position, the inputs that correspond to the switch packages connect to the same AND gate so that they are ANDed together.

- The CLOCK switch J_{10} is toggled to clock the D flip-flop. If the OUTPUT MODE switch J_{11} is closed (the macrocell is configured for combinational operation), the CLOCK switch has no effect.

- The OUTPUT MODE switch J_{11} determines whether the output of the OR gate connects directly to the output XOR gate (J_{11} closed) or through the D flip-flop (J_{11} open) for combinational or registered operation, respectively.

- The OUTPUT INVERT switch J_{12} determines whether the logic level from the OR gate is unchanged (J_{12} closed) or inverted (J_{12} open).

- The UNUSED SWITCH DISABLE switch J_{13} prevents any AND gates that are not used in the logic from affecting the output of the OR gate. Note that a device programmer would automatically program a real PLD device to ensure this.

2) Program the macrocell to implement the 2-input XOR combinational function. The SOP expression for a 2-input XOR function is $Y = A \cdot B' + A' \cdot B$. To implement this logic, do the following:

   a. Connect A and B' to AND gate U_1 by sliding the first switch of J_4 and J_7 UP.

   b. Connect A' and B to AND gate U_2 by sliding the second switch of J_5 and J_6 UP.

   c. Disable unused AND gates U_3 and U_4 by sliding the third and fourth switch of J_{13} UP.

   d. Close the OUTPUT MODE switch J_{11}.

   e. Close the OUTPUT INVERT switch J_{12}.

3) Start the simulation.

4) Use switches J_1, J_2, and J_3 to set inputs A, B, and C to the specified logic levels in Table 12-1. Record the resulting Y output.

Table 12-1: XOR Function Logic Levels

C	B	A	Y	C	B	A	Y
0	0	0		1	0	0	
0	0	1		1	0	1	
0	1	0		1	1	0	
0	1	1		1	1	1	

Are the outputs in Table 12-1 consistent with those for $Y = A$ XOR B?

5) Open the OUTPUT MODE switch J_{11} and repeat step 4. What happens?

6) Repeat step 4 but toggle the CLOCK switch J_{10} after each setting. What happens?

7) Stop the simulation.

12.6 Programming the 2-of-3 Voting Function

1) Open the Multisim file *Digital_Exp_12_Part_01*.

2) Record in Table 12-2 the necessary switch settings to implement the inverted combinational 2-of-3 voting function $Y' = CBA + CBA' + CB'A + C'BA$. The output Y for this function should be LOW whenever at least two inputs are HIGH. The first column, corresponding the first term, is done for you as an example.

Table 12-2: 2-of-3 Voting Function Switch Settings

AND Gate	U_1	U_2	U_3	U_4
Programmed Term	ABC	A'BC	AB'C	ABC'
J_4 (A)	UP			
J_5 (A')	DOWN			
J_6 (B)	UP			
J_7 (B')	DOWN			
J_8 (C)	UP			
J_9 (C')	DOWN			
J_{13} (UNUSED SWITCH DISABLE)	DOWN			
J_{11} (OUTPUT MODE)				
J_{12} (OUTPUT INVERT)				

3) Use your settings in Table 12-2 to program the macrocell.

4) Start the simulation.

5) Use switches J_1, J_2, and J_3 to set the A, B, and C inputs to the specified logic levels in Table 12-3. Record the resulting Y output.

Table 12-3: 2-of-3 Function Logic Levels

C	B	A	Y	C	B	A	Y
0	0	0		1	0	0	
0	0	1		1	0	1	
0	1	0		1	1	0	
0	1	1		1	1	1	

Is the output Y in Table 12-3 "0" when at least two inputs are "1"?

6) Stop the simulation.

Conclusions for Part 1

Questions for Part 1

1) How could you modify the macrocell programming to implement the XNOR function rather than the XOR function?

2) What fault would you suspect if you programmed the macrocell for registered operation but the output changed as soon as the inputs changed?

3) If you wished to simplify the 2-of-3 voting logic to $Y' = AB + AC + BC$, how would the programming change for J_4 through J_9 and J_{13}?

J_4:

J_5:

J_6:

J_7:

J_8:

J_9:

J_{13}:

Part 2:   The ROM Sum-of-Products Look-Up Table

12.7 Programming a 2-Input XOR Function

1) Open the Multisim file *Digital_Exp_12_Part_02*. The circuit simulates a 3-input SOP look-up table. The look-up table is a ROM implementation, as the AND terms are fixed and the OR term is programmable.

   - The eight AND gates U_1 through U_8 decode each of the possible 3-variable terms.
   - The switch pack J_4 selectively connects the decoded terms to the OR gate U_9 to create the desired SOP expression.

2) Program the look-up table to implement the 2-input XOR function $Y = A \cdot B' + A' \cdot B$. To implement the logic, do the following:

   a. Move Switch 2 (the switch for U_2) to the right to connect the $AB'C'$ term to the OR gate.

   b. Move Switch 6 (the switch for U_6) to the right to connect the $AB'C$ term to the OR gate.

   c. Move Switch 3 (the switch for U_3) to the right to connect the $A'BC'$ term to the OR gate.

   d. Move Switch 7 (the switch for U_7) to the right to connect the $A'BC$ term to the OR gate.

3) Start the simulation.

4) Use switches J_1, J_2, and J_3 to set inputs A, B, and C to the specified logic levels in Table 12-4. Record the resulting Y output.

Table 12-4: Observed XOR Function Logic Levels

C	B	A	Y	C	B	A	Y
0	0	0	0	1	0	0	0
0	0	1	1	1	0	1	1
0	1	0	1	1	1	0	1
0	1	1	0	1	1	1	0

Are the outputs in Table 12-1 consistent with those for $Y = A$ XOR B?

5) Stop the simulation.

12.8 Programming the 2-of-3 Voting Function

1) Open the Multisim file *Digital_Exp_12_Part_02*.

2) Record in Table 12-5 the necessary J_4 switch settings (LEFT or RIGHT) to implement the combinational 2-of-3 voting function $Y = CBA + CBA' + CB'A + C'BA$. The output Y for this function should be HIGH whenever at least two inputs are HIGH.

Table 12-5: 2-of-3 Voting Function J_4 Switch Settings

$A'B'C'$	$AB'C'$	$A'BC'$	ABC'	$A'B'C$	$AB'C$	$A'BC$	ABC
Switch 1	Switch 2	Switch 3	Switch 4	Switch 5	Switch 6	Switch 7	Switch 8

3) Use your settings in Table 12-5 to program the look-up table.

4) Start the simulation.

5) Use switches J_1, J_2, and J_3 to set the A, B, and C inputs to the specified logic levels in Table 12-6. Record the resulting Y output.

Table 12-6: 2-of-3 Function Logic Levels

C	B	A	Y	C	B	A	Y
0	0	0		1	0	0	
0	0	1		1	0	1	
0	1	0		1	1	0	
0	1	1		1	1	1	

Is the output Y in Table 12-3 "1" when at least two inputs are "1"?

6) Stop the simulation.

Conclusions for Part 2

Questions for Part 2

1) If you wished to expand the circuit of *Digital_Exp_12_Part_02* to six inputs, what changes would you need to make?

2) If you wished to simplify the 2-of-3 voting logic to $Y = AB + AC + BC$, can you change the programming for J_4?

Digital Experiment 13 - Signal Interfacing

13.1 Introduction

The earliest digital computers were primarily number-crunching behemoths that had little direct interaction with the real world. Today microprocessors are found in nearly every type of consumer product: automobiles, phones, cameras, music and video players, toys, household appliances, and many other everyday items. The circuits that allow digital technology to interact with an analog world are signal converters. Analog-to-digital converters (ADCs) convert continuously-varying analog signals into discrete digital data, and digital-to-analog converters (DACs) convert digital data back into analog signals. A digital scale is an example of an ADC application, as the scale converts an analog quantity (weight) to a digital value (the scale reading). Similarly, a CD player is an example of a DAC application, as digital data (the CD content) is converted back into an analog signal (sound waves).

The basic purpose of a DAC is to develop an analog signal that is proportional to an applied digital value. One way is for the DAC to compare the input value to a reference digital value and use this to develop a voltage based on the reference analog value. Another way is for the DAC to develop a proportional voltage (or current) for each bit position, and sum the resulting voltages (or currents) for the digital value.

The basic purpose of an ADC is to generate a digital value that is proportional to an applied analog signal. The ADC does this by repeatedly generating digital values, applying the value to a DAC, comparing the DAC output to the analog input, and using the result of the comparison to adjust the digital value. The various types of ADCs differ primarily in how they attempt to optimize adjusting the digital value and minimize the time to generate the final value.

In Part 1 of this experiment, you will examine the operations of a binary-weighted DAC and a single-ramp ADC. In Part 2, you will simulate a circuit that consists of an ADC combined with a DAC and compare the reconstructed analog waveform with the original analog waveform.

13.2 Reading

Floyd, *Digital Fundamentals, 10th Edition*, Chapter 12

13.3 Key Objectives

Part 1: Analyze and simulate the operations of a binary-weighted DAC and single-ramp ADC.

Part 2: Simulate a combined ADC and DAC circuit, compare the reconstructed analog waveform with the original analog waveform, and determine the effect of the sampling rate to the input frequency.

13.4 Multisim Files

Part 1: *Digital_Exp_13_Part_01a*, *Digital_Exp_13_Part_01b*, and *Digital_Exp_13_Part_01c*.

Part 2: *Digital_Exp_13_Part_02*.

Part 1: Converting Between Digital and Analog

13.5 The Binary-Weighted DAC

The binary-weighted DAC uses an op-amp summing circuit (which you will study in Devices Experiment 12) to develop an analog output proportional to the digital input. The output of an op-amp summing circuit is equal to

$$V_{OUT} = -[(V_{IN1} / R_{IN1}) + (V_{IN2} / R_{IN2}) + (V_{IN3} / R_{IN3}) + (V_{IN4} / R_{IN4})] R_F$$

where V_{IN1}, V_{IN2}, V_{IN3}, and V_{IN4} are the values of the four digital input voltages, R_{IN1}, R_{IN2}, R_{IN3}, and R_{IN4} are values of the input resistors to which the voltages are applied, and R_F is the value of the op-amp feedback resistor. Because the output of the summing amplifier is negative when input voltages are positive, the DAC adds an inverting amplifier that changes the negative voltage back to a positive voltage, so that

$$V_{OUT} = +[(V_{IN1} / R_{IN1}) + (V_{IN2} / R_{IN2}) + (V_{IN3} / R_{IN3}) + (V_{IN4} / R_{IN4})] R_F$$

1) For each of the values in Table 13-1, calculate and record to three significant digits the output voltage V_{OUT} for the binary weighted DAC.

Table 13-1: Calculated Binary Weighted DAC Output Voltage

V_{IN4}	V_{IN3}	V_{IN2}	V_{IN1}	R_{IN4}	R_{IN3}	R_{IN2}	R_{IN1}	R_F	V_{OUT}
0 V	0 V	0 V	0 V	250 kΩ	500 kΩ	1 MΩ	2 MΩ	125 kΩ	
0 V	0 V	0 V	+5 V	250 kΩ	500 kΩ	1 MΩ	2 MΩ	125 kΩ	
0 V	0 V	+5 V	0 V	250 kΩ	500 kΩ	1 MΩ	2 MΩ	125 kΩ	
0 V	0 V	+5 V	+5 V	250 kΩ	500 kΩ	1 MΩ	2 MΩ	125 kΩ	
0 V	+5 V	0 V	0 V	250 kΩ	500 kΩ	1 MΩ	2 MΩ	125 kΩ	
0 V	+5 V	0 V	+5 V	250 kΩ	500 kΩ	1 MΩ	2 MΩ	125 kΩ	
0 V	+5 V	+5 V	0 V	250 kΩ	500 kΩ	1 MΩ	2 MΩ	125 kΩ	
0 V	+5 V	+5 V	+5 V	250 kΩ	500 kΩ	1 MΩ	2 MΩ	125 kΩ	
+5 V	0 V	0 V	0 V	250 kΩ	500 kΩ	1 MΩ	2 MΩ	125 kΩ	
+5 V	0 V	0 V	+5 V	250 kΩ	500 kΩ	1 MΩ	2 MΩ	125 kΩ	
+5 V	0 V	+5 V	0 V	250 kΩ	500 kΩ	1 MΩ	2 MΩ	125 kΩ	
+5 V	0 V	+5 V	+5 V	250 kΩ	500 kΩ	1 MΩ	2 MΩ	125 kΩ	
+5 V	+5 V	0 V	0 V	250 kΩ	500 kΩ	1 MΩ	2 MΩ	125 kΩ	
+5 V	+5 V	0 V	+5 V	250 kΩ	500 kΩ	1 MΩ	2 MΩ	125 kΩ	
+5 V	+5 V	+5 V	0 V	250 kΩ	500 kΩ	1 MΩ	2 MΩ	125 kΩ	
+5 V	+5 V	+5 V	+5 V	250 kΩ	500 kΩ	1 MΩ	2 MΩ	125 kΩ	

2) Open the Multisim file *Digital_Exp_13_Part_01a*.

3) Double-click the multimeter XMM_1 to expand it and set it for dc voltmeter mode.

4) Start the simulation.

5) For each of the rows in Table 13-2:

    a. Set the switches for J_1 to apply the specified input voltages.

    b. Record to three significant digits the measured voltage as V_{OUT}.

Table 13-2: Measured Binary Weighted DAC Output Voltage

V_{IN4}	V_{IN3}	V_{IN2}	V_{IN1}	V_{OUT}	V_{IN4}	V_{IN3}	V_{IN2}	V_{IN1}	V_{OUT}
0 V	0 V	0 V	0 V		+5 V	0 V	0 V	0 V	
0 V	0 V	0 V	+5 V		+5 V	0 V	0 V	+5 V	
0 V	0 V	+5 V	0 V		+5 V	0 V	+5 V	0 V	
0 V	0 V	+5 V	+5 V		+5 V	0 V	+5 V	+5 V	
0 V	+5 V	0 V	0 V		+5 V	+5 V	0 V	0 V	
0 V	+5 V	0 V	+5 V		+5 V	+5 V	0 V	+5 V	
0 V	+5 V	+5 V	0 V		+5 V	+5 V	+5 V	0 V	
0 V	+5 V	+5 V	+5 V		+5 V	+5 V	+5 V	+5 V	

6) Stop the simulation.

Observations:

7) Open the Multisim file *Digital_Exp_13_Part_01b*. This circuit is identical to the circuit in *Digital_Exp_13_Part_01a*, but uses the word generator to apply the sequential digital values 0000 to 1111 to the DAC and the oscilloscope to observe the analog output.

8) Double-click the word generator and oscilloscope to expand them.

9) Click the **Cycle** button on the word generator to apply the digital sequence to the DAC. Allow the word generator to cycle at least twice so that the oscilloscope displays the complete pattern.

10) Stop the simulation.

Observations:

13.6 The Single-Ramp ADC

The single-ramp ADC uses a binary counter to apply digital values to a DAC (the binary-weighted DAC you already examined). An op-amp comparator compared the analog output of the DAC to the input voltage. As long as the DAC output is less than the input voltages, the comparator output is LOW, allowing the counter to run. As soon as the DAC output is greater than the input voltage, the comparator output goes HIGH, latching the counter value and resetting the circuit so that the conversion process begins again.

1) Open the Multisim file *Digital_Exp_13_Part_01c*.

2) Double-click the multimeter XMM_1 and set it to the dc voltmeter mode.

3) Start the simulation.

4) Press the "R" key to open and close switch J_2. This will reset the circuit and start the ADC conversions.

5) For each of the rows in Table 13-3:

   a. Use "Shift + V" to adjust potentiometer R_{10} to set the wiper to the indicated setting. Alternatively, you can hover the cursor over the potentiometer so that the slider appears and use the cursor to adjust the slider to "0%".

   b. Record to three significant digits the measured value of V_{IN} on the multimeter.

   c. Record the digitized value of V_{IN} on the logic probes X_6 through X_9.

   d. Divide V_{IN} by the digitized value. Record it to three significant digits as the Volts / Count value.

Table 13-3: Measured Single-Ramp ADC Values

Wiper Setting	V_{IN}	Digitized Value	Volts / Count	Wiper Setting	V_{IN}	Digitized Value	Volts / Count
0%				60%			
10%				70%			
20%				80%			
30%				90%			
40%				100%			
50%							

6) Stop the simulation.

Observations:

Conclusions for Part 1

Questions for Part 1

1) How could you increase the gain of the DAC circuit in *Digital_Exp_13_Part_01a* so that the output voltage for each digital input is twice that of the original circuit?

2) What fault would you suspect if the DAC circuit of *Digital_Exp_13_Part_01b* produced the waveform in Figure 13-1?

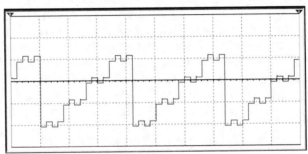

Figure 13-1: Faulty DAC Output

Part 2:   Signal Digitization

1) Open the Multisim file *Digital_Exp_13_Part_02*.
2) Double-click the oscilloscope to expand it.
3) Start the simulation.
4) Press the "R" key to open and close switch J_2. This will reset the circuit and begin digitizing the input waveform.
5) Allow the circuit to run for at least one period of the input waveform. Depending upon the speed of your computer, this may take a while. How well does the digitized waveform follow the original waveform?

6) Stop the simulation.
7) Double-click on the sampling clock source V_S to open its properties window.
8) Change the **Pulse Width:** value to "50 μsec" and the **Period:** value to "100 μsec".
9) Start the simulation.
10) Press the "R" key to open and close switch J_2 to reset the circuit and begin digitizing the input waveform.

11) Allow the circuit to run for at least one period of the input waveform. Depending upon the speed of your computer, this may take a while. How does the simulation speed compare with the original digitization? How well does the digitized waveform follow the original waveform?

12) Change the **Pulse Width:** value to "500 µsec" and the **Period:** value to "1 msec".

13) Start the simulation.

14) Press the "R" key to open and close switch J_2 to reset the circuit and begin digitizing the input waveform.

15) Allow the circuit to run for at least one period of the input waveform. How well do the shape and frequency of the digitized waveform compare with the original waveform?

16) Change the **Pulse Width:** value to "5 msec" and the **Period:** value to "10 msec".

17) Start the simulation.

18) Press the "R" key to open and close switch J_2 to reset the circuit and begin digitizing the input waveform.

19) Allow the circuit to run for at least one period of the input waveform. What is the shape of the digitized waveform?

20) Stop the simulation.

Conclusions for Part 2

Questions for Part 2

1) How could you modify the circuit in *Digital_Exp_13_Part_02* so that the digitized waveform more closely resembled the input waveform without increasing the sampling frequency?

2) What fault would you suspect if the circuit of *Digital_Exp_13_Part_02* produced the waveform in Figure 13-2?

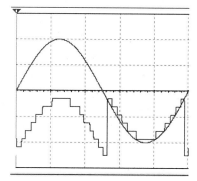

Figure 13-2: Faulty Digitized Signal

Digital Experiment 14 - Processor Support Circuitry

14.1 Introduction

Microprocessors (commonly called "processors") must interact with almost every part of the computer system. Processors must transfer data to and from memory, input/output (I/O) devices, peripherals, and other components, and respond to requests from these components to do so. The processor sends and receives information over system buses, or groups of conductors, or signal lines, carrying signals of the same nature. The three main system buses are the *address bus* with which the processor accesses specific devices, the *data bus* over which information passes to and from the processor, and the *control bus* which coordinates the operation of the computer system.

Many of the bus lines are shared by several devices. In particular, the data bus must allow the processor and other devices to drive the data lines HIGH or LOW when they must send data over the bus to another device. A device must not only avoid sending data at the same time as another, but also avoid holding the data lines HIGH or LOW when the device is inactive. If the outputs of more than one device are active on a bus at the same time, valid data on the bus can be corrupted and the devices can damage each other. The processor and control bus ensure that only one device is active and using the data bus at any one time. Tri-state outputs (also called high-impedance or high-Z outputs) essentially disconnect inactive devices from the data bus so that they do not attempt to drive the bus. Experiment 11 used tri-state buffers to allow multiple memory cells to share the same data line. Only the memory cells being read were able to drive the data lines. Many devices that must connect to the data bus in computer systems have tri-state outputs, but not all. If devices must share signal lines but do not have tri-state outputs, the circuit design must add tri-state buffers.

To ensure that only one device at a time is active on the data bus, the processor uses the control bus to supervise bus access. Usually devices, such as memory or a hard drive, send data over the data bus at the request of the processor, but some devices have data, such as a keypress or an e-mail message, that the processor does not specifically request. A way to handle this is for the processor to periodically poll, or check, each device in the system, and allow those that have data to send it. Polling is very simple, but very inefficient as the processor spends time checking for data even if no data is present.

A much more efficient way is for devices to signal the processor that they have information to send and for the processor to respond when it receives a signal. Devices in a computer system signal the processor by generating what is called an interrupt request (IRQ). An IRQ causes the processor to temporarily suspend what it is doing, generate an interrupt acknowledge (IACK), and run special code called an interrupt service routine (ISR) to deal with the interrupt. Once the processor has serviced the IRQ, it resumes the task from which it was interrupted. To assist the processor in handling interrupts, a computer system will usually include a programmable interrupt controller (PIC) to ensure that interrupts are prioritized and serviced in an orderly manner. Interrupt structures can be pre-emptive or non-preemptive. In systems with preemptive interrupt structures, processors will suspend servicing an interrupt if a higher-priority interrupt occurs. In systems with non-preemptive interrupt structures, processors will complete servicing an interrupt before recognizing any new interrupts.

In Part 1 of this experiment, you will demonstrate and observe the nature and operation of tri-state buffers. In Part 2, you will investigate how a priority interrupt circuit operates.

14.2 Reading

Floyd, *Digital Fundamentals, 10th Edition*, Chapter 13.

14.3 Key Objectives

Part 1:   Observe and analyze the operation of tri-state buffers on shared signal lines.

Part 2:   Simulate the operation of a typical priority interrupt controller circuit.

14.4 Multisim Circuits

Part 1:   *Digital_Exp_14_Part_01a* and *Digital_Exp_14_Part_01b*

Part 2:   *Digital_Exp_14_Part_02*

Part 1:   Operation of Tri-State Buffers

1) Open the Multisim file *Digital_Exp_14_Part_01a*. U_1 and U_2 are the tri-state buffers. In addition to an input and output pin, each buffer has an enable line that is active-LOW (indicated by the inversion bubble). When the enable line is LOW, the output is at the same logic level as the input. When the enable line is HIGH, the output is tri-stated so that the output appears as a high impedance and is effectively disconnected from the shared output line.

2) Start the simulation.

3) For each row in Table 14-1:

   a. Set each of the switches to the indicated positions.

   b. Record the states of the logic probes for Q_0 through Q_3 (OFF = 0, ON = 1).

Table 14-1: Tri-State Buffer Circuit Switch Settings and Signal Levels

J_1				J_2				J_3 CLOSED				J_3 OPEN			
A_0	A_1	A_2	A_3	B_0	B_1	B_2	B_3	Q_0	Q_1	Q_2	Q_3	Q_0	Q_1	Q_2	Q_3
0	0	0	0	1	0	0	0								
0	0	0	0	0	1	0	0								
0	0	0	0	0	0	1	0								
0	0	0	0	0	0	0	1								
1	1	1	1	1	0	0	0								
1	1	1	1	0	1	0	0								
1	1	1	1	0	0	1	0								
1	1	1	1	0	0	0	1								
1	0	0	0	0	0	0	0								
0	1	0	0	0	0	0	0								
0	0	1	0	0	0	0	0								
0	0	0	1	0	0	0	0								
1	0	0	0	1	1	1	1								
0	1	0	0	1	1	1	1								
0	0	1	0	1	1	1	1								
0	0	0	1	1	1	1	1								

4) Stop the simulation.

   Observations:

5) Delete the wire segment between the Y_1' of the 2-to-4 decoder U_{3A} and the enable lines of U_{2A} through U_{2D}.

6) Use the **Place Source** tool to add a digital ground (DGND) to the circuit. Connect the enable lines of U_{2A} through U_{2D} to digital ground.

7) Start the simulation.

8) For each row in Table 14-2:

   a. Set each of the switches for the indicated positions.

   b. Record the states of the logic probes for Q_0 through Q_3.

Table 14-2: Tri-State Buffer Circuit Switch Settings and Signal Levels

J_1				J_2				J_3 CLOSED				J_3 OPEN			
A_0	A_1	A_2	A_3	B_0	B_1	B_2	B_3	Q_0	Q_1	Q_2	Q_3	Q_0	Q_1	Q_2	Q_3
0	0	0	0	1	0	0	0								
0	0	0	0	0	1	0	0								
0	0	0	0	0	0	1	0								
0	0	0	0	0	0	0	1								
1	1	1	1	1	0	0	0								
1	1	1	1	0	1	0	0								
1	1	1	1	0	0	1	0								
1	1	1	1	0	0	0	1								
1	0	0	0	0	0	0	0								
0	1	0	0	0	0	0	0								
0	0	1	0	0	0	0	0								
0	0	0	1	0	0	0	0								
1	0	0	0	1	1	1	1								
0	1	0	0	1	1	1	1								
0	0	1	0	1	1	1	1								
0	0	0	1	1	1	1	1								

9) Stop the simulation. Summarize your observations.

10) Open the Multisim file *Digital_Exp_14_Part_01b*. How does this circuit differ from that of *Digital_Exp_14_Part_01a*?

11) Does the operation of the circuit appear to be the same as that of *Digital_Exp_14_Part_01a*? If so, why might this still not be a good way to enable the tri-state buffers?

Conclusions for Part 1

Questions for Part 1

1) How could you modify the circuit of *Digital_Exp_14_Part_01a* to connect two more 4-position DIP switches to the signal lines?

2) Suppose that when you simulated the circuit of *Digital_Exp_14_Part_01a*, you found that Q_1 was always LOW when either A_1 or B_1 was LOW and J_3 was OPEN, but that the logic level of Q_1 was the same as that for A_1 when J_3 was open. What fault would you suspect?

Part 2:   Interrupt Controller Circuit

In this part of the experiment you will simulate the process by which a processor receives, acknowledges, and responds to interrupts. The process is as follows:

a.   One or more devices assert interrupt requests (IRQs).

b.   The processor reads the indicated interrupt level.

c.   The processor generates an interrupt acknowledge (IACK).

d.   The processor masks (ignores) any new interrupts while servicing the highest indicated interrupt. A Level 0 interrupt is the lowest and least important, and a Level 7 is the highest and most important.

e.   The serviced device negates its IRQ.

f.   The processor negates it IACK.

While the processor is servicing an interrupt, it is possible that more interrupts from other devices can occur. If so, when the processor negates its IACK, the processor will recognize and acknowledge the highest interrupt that has not yet been acknowledged or serviced.

1) Open the Multisim file *Digital_Exp_14_Part_02*.

2) Simulate a Level 2 interrupt by sliding the third switch from the top of J_1 to the right. Describe what happens.

3) Close switch J_2 to simulate an IACK from the processor. Describe what happens.

4) Simulate that a Level 0, Level 4, and Level 7 interrupt occur while the processor is servicing the Level 2 interrupt by sliding the corresponding switches of J_1 to the right. Describe what happens and why.

5) Simulate that the processor has serviced the Level 2 interrupt by sliding the Level 2 switch of J_1 to the left and opening J_2. Describe what happens.

6) Close switch J_2 to simulate an IACK from the processor. Describe what happens.

7) Simulate that the processor has serviced the Level 7 interrupt by sliding the Level 7 switch of J_1 to the left and opening J_2. Describe what happens and why.

8) Close switch J_2 to simulate an IACK from the processor. Describe what happens.

9) Simulate that the processor has serviced the Level 4 interrupt by sliding the Level 4 switch of J_1 to the left and opening J_2. Describe what happens and why.

10) Close switch J_2 to simulate an IACK from the processor. Describe what happens.

11) Simulate that a Level 3 interrupt occurs while the processor is servicing the Level 0 interrupt by sliding the corresponding switch of J_1 to the right. Describe what happens and why.

12) Simulate that the processor has serviced the Level 0 interrupt by sliding the Level 0 switch of J_1 to the left and opening J_2. Describe what happens.

13) Close switch J_2 to simulate an IACK from the processor. Describe what happens.

14) Simulate that the processor has serviced the Level 3 interrupt by sliding the Level 3 switch of J_1 to the left and opening J_2. Describe what happens and why.

Conclusions for Part 2

Questions for Part 2

1) From your observations, how does the circuit indicate that interrupts are masked and that the processor is servicing an active interrupt?

2) Suppose that the processor is servicing a Level 2 interrupt while Level 1 and Level 3 interrupts are pending, and that a Level 5 interrupt occurs before the processor finishes servicing the Level 2 interrupt. If no further interrupts occur, what is the order in which the processor will service the pending interrupts?

3) Is the circuit in *Digital_Exp_14_Part_02* preemptive or non-preemptive? Explain your answer.

Digital Experiment 15 - The Arithmetic Logic Unit

15.1 Introduction

The fundamental microprocessor (often simply called "the processor") consists of three functional units: the register array, the timing and control block, and the arithmetic logic unit (ALU). The register array provides temporary storage for the processor so that it can execute programs and process program data. The timing and control block is essentially a state machine that generates the necessary internal and external signals so that the processor can access instructions and program data, execute instructions, and process information in the right order and at the right time. The ALU contains the circuitry that allows the processor to perform calculations and manipulate data. Previous experiments introduced the concepts of data registers and state machines. In this experiment you will examine the organization and operation of the ALU.

In essence, an ALU contains a number of logic gates, shift registers, adders, and other functional logic blocks that allow it to perform a variety of logic and arithmetic operations. Note that this is not the same as a programmable logic device (PLD), which you studied in Experiment 12. A PLD is a device that can be programmed (internally configured) to implement a desired logic function. An ALU contains fixed logic functions that are accessed by applying a specific pattern to external select lines.

In Part 1 of this experiment, you will analyze and verify the operation of a small-scale, four-function ALU. In Part 2, you will investigate the 74LS181 ALU, which is an actual four-bit ALU with sixteen arithmetic and logic functions.

15.2 Reading

Floyd, *Digital Fundamentals, 10th Edition*, Chapter 13

15.3 Key Objectives

Part 1: Analyze the design of a simple ALU, determine the signals required to access a specific function, and verify the logic and arithmetic functions of the ALU.

Part 2: Interpret the signal specifications of the 74LS181 ALU, determine signals required to access a specific function, and verify the required signals for and operation of the specified function.

15.4 Multisim Files

Part 1: *Digital_Exp_15_Part_01*

Part 2: *Digital_Exp_15_Part_02*

Part 1:   Basic ALU Design and Operation

1) Open the Multisim file *Digital_Exp_15_Part_01*.

2) Analyze the decoder circuitry of U_1 to determine which logical function corresponds to each combination of S_0 and S_1, and record the function for each in Table 15-1. Note that a closed switch corresponds to a "0" and an open switch corresponds to a "1".

3) For each combination of A, B, and C_{IN} in Table 15-1, record the expected logic states ("0" or "1") for Y and C_{OUT} for each of your recorded functions.

Table 15-1: Predicted ALU Functions and Output Values

C_{IN}	A	B	$S_1S_0 = 00$ AND		$S_1S_0 = 01$ OR		$S_1S_0 = 10$ XOR		$S_1S_0 = 11$ FULL ADDER	
			Y	C_{OUT}	Y	C_{OUT}	Y	C_{OUT}	Y	C_{OUT}
0	0	0								
0	0	1								
0	1	0								
0	1	1								
1	0	0								
1	0	1								
1	1	0								
1	1	1								

4) For each row and column in Table 15-2, set switches J_1 through J_5 to apply the specified values of A, B, C_{IN}, S_0, and S_1 to the circuit and record the states of the logic probes for Y and C_{OUT}. Note that an "OFF" state corresponds to a "0" and "ON" corresponds to a "1".

Table 15-2: Observed ALU Functions and Output Values

C_{IN}	A	B	$S_1S_0 = 00$		$S_1S_0 = 01$		$S_1S_0 = 10$		$S_1S_0 = 11$	
			Y	C_{OUT}	Y	C_{OUT}	Y	C_{OUT}	Y	C_{OUT}
0	0	0								
0	0	1								
0	1	0								
0	1	1								
1	0	0								
1	0	1								
1	1	0								
1	1	1								

Observations:

Conclusions for Part 1

Questions for Part 1

1) What is the purpose of AND gates U_{2B} and U_{2C} in the ALU circuit?

2) Suppose that, in step 4, you find that the ALU circuit functions as an XOR gate when you set the switches for $S_1 S_0 = 00$ and $S_1 S_0 = 10$ and as a full adder when you set the switches for $S_1 S_0 = 01$ and $S_1 S_0 = 11$. What fault would you suspect?

3) How would you modify the ALU to include a total of four additional arithmetic and logic functions?

Part 2:   The 74LS181 ALU

The 74LS181 is a 24-pin TTL 4-bit ALU that incorporates 16 logic functions and 16 arithmetic functions that are selected by four function-select inputs (S_0 through S_3) and a mode control input (M). Table 15-3 is a partial summary of the function outputs for the various combinations of these input pins.

Table 15-3: 74LS181 4-Bit ALU Function Outputs

S_3	S_2	S_1	S_0	$M = 1$ (Logic Functions)	$M = 0$ (Arithmetic Functions)	
					$C_n' = 1$	$C_n' = 0$
0	0	0	0	A'	A	A plus 1
0	0	0	1	$(A + B)'$	$A + B$	$(A + B)$ plus 1
0	0	1	0	$A' \cdot B$	$A + B'$	$(A + B')$ plus 1
0	0	1	1	0000	1000	0000
0	1	0	0	$(A \cdot B)'$	A plus $(A \cdot B')$	A plus $(A \cdot B')$ plus 1
0	1	0	1	B'	$(A + B)$ plus $(A \cdot B')$	$(A + B)$ plus $(A \cdot B')$ plus 1
0	1	1	0	A XOR B	A minus B minus 1	A minus B
0	1	1	1	$A \cdot B'$	$(A \cdot B')$ minus 1	$A \cdot B'$
1	0	0	0	$A' + B$	A plus $(A \cdot B)$	A plus $(A \cdot B)$ plus 1
1	0	0	1	A XNOR B	A plus B	A plus B plus 1
1	0	1	0	B	$(A + B')$ plus $(A \cdot B)$	$(A \cdot B)'$ plus $(A \cdot B)$ plus 1
1	0	1	1	$A \cdot B$	$(A \cdot B)$ minus 1	$A \cdot B$
1	1	0	0	1111	A plus A	A plus A plus 1
1	1	0	1	$A + B'$	$(A + B)$ plus A	$(A + B)$ plus A plus 1
1	1	1	0	$A + B$	$(A + B')$ plus A	$(A + B')$ plus A plus 1
1	1	1	1	A	A minus 1	A

In addition to the 4-bit A and B inputs and the 4-bit function output, the ALU also includes a C_n' carry input, C_{n+4}' carry output, a G' carry generate output, and a P' carry propagate output for look-ahead carry adder applications, and an $A=B$' comparator output. Note that all of these signals are active-LOW. For example, C_{in}' must be LOW to indicate a carry in, and $A=B$' will be HIGH if A is not equal to B.

15.5 Basic Logic Functions

1) From Table 15-3, identify and record the ALU inputs to implement the logic functions in Table 15-4. The first row is done for you as an example.

Table 15-4: 74LS181 ALU Inputs for Basic Logic Functions

Function	M	S_3	S_2	S_1	S_0	C_n'
A'	1	0	0	0	0	X
B'						
$A \cdot B$						
$A + B$						
A XOR B						
A XNOR B						

2) For the specified values of A and B for each row in Table 15-5, determine and record the 4-bit function output for the function specified in each column. The first two rows are done for you as an example.

Table 15-5: Calculated Logic Function Outputs

A	B	A'	B'	$A \cdot B$	$A + B$	A XOR B	A XNOR B
0000	0000	1111	1111	0000	0000	0000	1111
0000	0101	1111	1010	0000	0101	0101	1010
0000	1010						
0000	1111						
0101	0000						
0101	0101						
0101	1010						
0101	1111						
1010	0000						
1010	0101						
1010	1010						
1010	1111						
1111	0000						
1111	0101						
1111	1010						
1111	1111						

3) Open the Multisim file *Digital_Exp_15_Part_02*.

4) Start the simulation.

5) For each of the columns in Table 15-6:
   a. Use your values in Table 15-4 to set the "FUNCTION" and "MODE" switches to implement the indicated logic function.
   b. Set the "INPUT A" and "INPUT B" switches to each of the specified values of A and B.
   c. Record the $F_3F_2F_1F_0$ function output values indicated by the logic probes.
6) Stop the simulation.

Table 15-6: Observed 74LS181 Function Outputs

A	B	A'	B'	$A \cdot B$	$A + B$	A XOR B	A XNOR B
0000	0000						
0000	0101						
0000	1010						
0000	1111						
0101	0000						
0101	0101						
0101	1010						
0101	1111						
1010	0000						
1010	0101						
1010	1010						
1010	1111						
1111	0000						
1111	0101						
1111	1010						
1111	1111						

Observations:

15.6 Basic Arithmetic Functions

1) From Table 15-3, identify and record the ALU inputs to implement the arithmetic functions in Table 15-7.

Table 15-7: 74LS181 ALU Inputs for Basic Arithmetic Functions

Function	M	S_3	S_2	S_1	S_0	C_n'
A plus B						
A minus B						

2) For the specified values of A and B for each row in Table 15-8, determine and record the 4-bit function and carry outputs for the arithmetic operation specified in each column. For the "A minus B" operation, the ALU adds the 2's complement of B to A. $C_{OUT}' = 0$ if the answer is positive (the addition generates a carry) and $C_{OUT}' = 1$ if the answer is negative and in 2's complement form (the addition generates no carry).

Table 15-8: Calculated Arithmetic Function Outputs

A	B	A plus B		A minus B	
		C_{OUT}'	$F_3F_2F_1F_0$	C_{OUT}'	$F_3F_2F_1F_0$
0000	0000				
0000	0101				
0000	1010				
0000	1111				
0101	0000				
0101	0101				
0101	1010				
0101	1111				
1010	0000				
1010	0101				
1010	1010				
1010	1111				
1111	0000				
1111	0101				
1111	1010				
1111	1111				

3) Open the Multisim file *Digital_Exp_15_Part_02*.

4) Start the simulation.

5) For each of the columns in Table 15-9:

    a.   Use your values in Table 15-7 to set the "FUNCTION" and "MODE" switches to implement the indicated arithmetic function.

    b.   Set the "INPUT A" and "INPUT B" switches to each of the specified values of *A* and *B*.

    c.   Record the C_{OUT}' and $F_3F_2F_1F_0$ function output values of the logic probes.

6) Stop the simulation.

Table 15-9: Observed Arithmetic Function Outputs (continued on next page)

A	B	A plus B		A minus B	
		C_{OUT}'	$F_3F_2F_1F_0$	C_{OUT}'	$F_3F_2F_1F_0$
0000	0000				
0000	0101				
0000	1010				
0000	1111				
0101	0000				
0101	0101				

A	B	A plus B		A minus B	
		C_{OUT}'	$F_3F_2F_1F_0$	C_{OUT}'	$F_3F_2F_1F_0$
0101	1010				
0101	1111				
1010	0000				
1010	0101				
1010	1010				
1010	1111				
1111	0000				
1111	0101				
1111	1010				
1111	1111				

Observations:

Conclusions for Part 2

Questions for Part 2

1) What are three ways that you could use the 74LS181 to invert a binary number?

2) What are two ways that you could use the 74LS181 to increment a binary number by 1?

3) Each of the 74LS181 verification tests used a limited number of values of A and B for each of the specified logic and arithmetic functions. How many test cases would each function require to fully test the four-bit ALU?

Electronic Devices Experiments

Devices Experiment 1 - Semiconductor Diodes **213**

Devices Experiment 2 - Diode Applications **221**

Devices Experiment 3 - Bipolar Junction Transistors **229**

Devices Experiment 4 - BJT Biasing..**237**

Devices Experiment 5 - BJT Amplifiers **243**

Devices Experiment 6 - Power Amplifiers...**249**

Devices Experiment 7 - JFET Characteristics **255**

Devices Experiment 8 - FET Amplifiers **261**

Devices Experiment 9 - Amplifier Frequency Response....................................**267**

Devices Experiment 10 - Thyristors...**273**

Devices Experiment 11 - The Operational Amplifier...**277**

Devices Experiment 12 - Basic Op-Amp Circuits...**283**

Devices Experiment 13 - Special-Purpose Op-Amps **289**

Devices Experiment 14 - Active Filters..**297**

Devices Experiment 15 - Oscillators **305**

Devices Experiment 1 - Semiconductor Diodes

1.1 Introduction

Semiconductor diodes are created by the joining together of *p*- and *n*-materials at the time of manufacturing, creating a one-way valve for current. The region near the joined area is called the *depletion region*. If an external voltage is applied to the diode, it is said to be biased. Reverse-bias is the condition where a positive voltage is connected to the *n*-region and a negative voltage is applied to the *p*-region. This creates a widening of the depletion region and effectively reduces the current to zero. Forward-bias is the condition where a positive voltage is connected to the *p*-region and a negative voltage is applied to the *n*-region. This creates a narrower depletion region and allows charge carriers to cross, thus enabling current.

In Part 1 of this experiment, you will use Multisim to simulate a diode test, take data for the diode characteristic and plot the response. In Part 2, you will observe the characteristics with a simulated oscilloscope and obtain the plot of other types of diodes. Part 2 lists steps for setting up the oscilloscope for this.

1.2 Reading

Floyd, *Electronic Devices, 8th Edition*, Chapter 1

1.3 Key Objectives

Part 1:   Measure the forward and reverse characteristics of a signal diode and observe the relationship between current and voltage on a semilog plot.

Part 2:   Plot the forward and reverse characteristics of a signal diode, a zener diode, and an LED using the oscilloscope and compare the forward ac resistance of each.

1.4 Multisim Files

Part 1:   *Devices Exp_01_Part_01*

Part 2:   *Devices Exp_01_Part_02*

Part 1:   The Diode Characteristic Curve

1.5 Diode Test with a Meter

A quick diode test is useful if you are not sure if a diode is good. Digital meters usually have a diode test position, which allows the meter to provide bias voltage for the diode. Figure 1-1 shows a meter with the diode test position.

1) Open the Multisim file *Devices Exp_01_Part_01*. This circuit simulates the reading you would see if the DMM in Multisim had a diode test position. The source represents a single 1.5 V battery inside the meter. Notice that the DMM shows the forward drop on the diode (in this case about 0.6 V). In practical meters, the reverse position will show an overload "OL" indication, but if you reverse the multimeter leads the simulation will not show this.

2) Keep the circuit file open. You will use it to complete the remaining sections of this experiment.

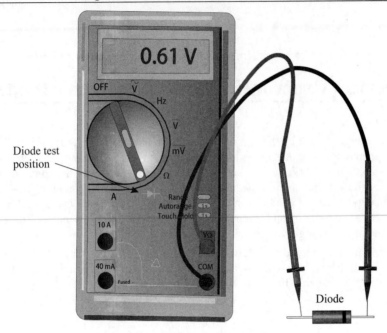

Figure 1-1: Typical Portable DMM

1.6   Forward Characteristic Curve

1) Start the simulation.

2) Set the source voltage to each value given in Table 1-1. Placing a voltmeter across R_1 is a common method used to determine current in actual circuits because it does not require you to break the circuit, as you would to insert an ammeter.

3) Read the voltages V_{R1} and V_{D1}. Record them in Table 1-1.

4) Stop the simulation.

5) Use Ohm's law to calculate and record to three significant digits the current I through the resistor.

6) Repeat Steps 1 through 5 for each value of V_S shown in Table 1-1.

Table 1-1: Current and Voltage for a Simulated 1N914 Diode

V_S	V_{R1}	V_{D1}	I
0.50 V			
0.75 V			
1.0 V			
1.5 V			
2.0 V			
4.0 V			
8.0 V			
12.0 V			
16.0 V			

7) Plot the *V-I* characteristic curve for the diode in Plot 1-1 by plotting the diode voltage V_{D1} on the *x*-axis and the diode current *I* on the *y*-axis.

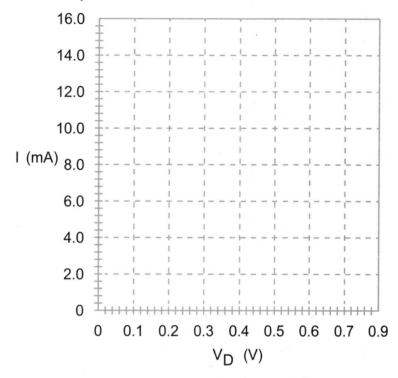

Plot 1-1: Forward *V-I* Characteristic Curve

8) Plot the same data on the semi-logarithmic plot in Plot 1-2. This type of plot will show details of the data when it is close to the *x*-axis.

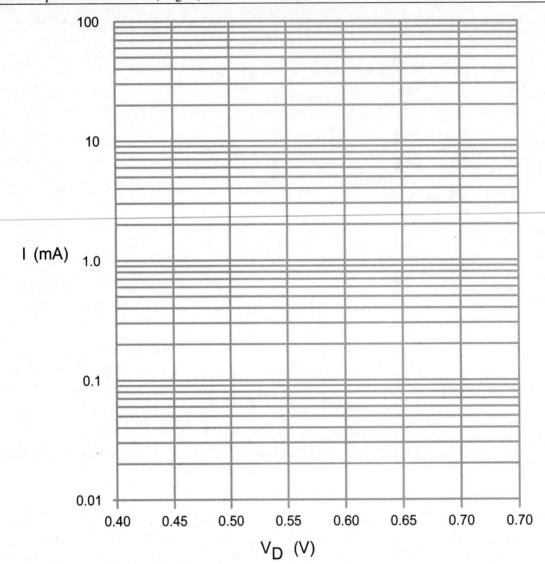

Plot 1-2: Forward V-I Semi-log Characteristic Curve

1.7 Reverse-Biasing the Diode

1) Change the value of R_1 to 1.0 MΩ.

2) Disconnect the wires from D_1.

3) Reverse the polarity of the diode.

4) Reconnect D_1 to the circuit.

5) Start the simulation.

6) Set the voltage for 10 V.

7) Measure the voltage across R_1.

8) Stop the simulation.

9) Use Ohm's law to calculate the current. In a circuit like this, most practical ammeters are not sensitive enough to measure the tiny current, so the measurement is often done indirectly using a voltmeter.

10) The measurement of current is subject to loading error if the voltmeter's internal resistance is not sufficiently high (at least 10 MΩ). You can check the internal resistance of a voltmeter by measuring a

known power supply voltage through a series 1.0 MΩ resistor. If the meter reads the supply voltage accurately, it has high input impedance.

Notice that the reverse-biased diode appears to be almost an open circuit in the simulation.

Observations:

Conclusions for Part 1

Questions for Part 1

1) Using the data taken in Table 1-1, what is the maximum power dissipated in the resistor?

$P_{R1} = VI =$

2) Using the data taken in Table 1-1, what is the maximum power dissipated in the diode?

$P_{D1} = VI =$

3) The data plotted on semi-log paper in Plot 1-2 is nearly a straight line. What conclusion can you draw from this fact?

4) In Step 1 of Section 1.7, you were directed to change the resistor to 1.0 MΩ to measure the reverse current. What advantage is this for the indirect measurement of reverse current in a practical circuit?

5) Assume that when you measure a power supply voltage through a series 1 MΩ resistor the voltage drops by half compared to the direct measurement. What is the input resistance of the meter? Explain your answer.

Part 2:   Plotting Diode Curves with an Oscilloscope

To directly view diode curves, you will use the Tektronix oscilloscope in *X-Y* mode, in which Channel 1 controls the *x*-axis and Channel 2 controls the *y*-axis. Channel 1 will sense the voltage drop across the diode and Channel 2 will sense the voltage across a 1.0 kΩ resistor, which represents a signal that is proportional to the current. Because the Channel 2 voltage is across a 1.0 kΩ resistor, you can convert the VOLTS/DIV control to a CURRENT/DIV control by dividing the setting by 1.0 kΩ.

1.8   Small-Signal Diode *V-I* Curve

1) Open the file *Devices_Exp_01_Part_02*.
2) Double-click the Tektronix oscilloscope and turn it on.
3) From the **DISPLAY** menu, choose **Format XY**.
4) From the **CH 2 MENU**, choose **Invert ON**.

5) Set the **CH 1 VOLTS/DIV** control to "200 mV".

6) Set the **CH 2 VOLTS/DIV** control to "2 V".

7) Use the **CH 2 POSITION** control to move the ground to the bottom graticule line. To do this accurately, you can temporarily select **Ground** for **CH 2 Coupling** and then return the **CH 2 Coupling** to **DC**. You should see the *V-I* curve for the diode on the display.

8) Use the **CH 1 Coupling** and **CH 1 POSITION** controls to verify that the center vertical axis represents 0 V.

9) Select **Ground** for **CH 1 Coupling**.

10) Use the **CH 1 POSITION** control to move the line to center screen.

11) Return the **CH 1 Coupling** to **DC**.

12) At this point you can check a point or two on the *V-I* curve and confirm that the plot is equivalent to the linear plot in Part 1. Describe a specific measured point on the curve and what it represents.

1.9 Other Diode *V-I* Curves

1) Turn off the power and replace the diode with a 1N4001GP, which is a rectifier diode. You will need to orient the diode so that the cathode is up (i.e., the diode arrow "points up"). Reconnect the circuit and turn the power on.

Observations for 1N4001GP:

2) Turn off the power and replace the 1N4001GP with a red LED. Change the **CH 1 VOLTS/DIV** control to 500 mV/div and observe the *V-I* curve. Turn the power on again.

Observations for Red LED:

3) Turn off the power and replace the red LED with a 1N4733A zener diode. A zener diode is a special diode designed to break down when reverse-biased, so you must orient the diode so that the cathode is down (i.e., the diode arrow points down).

4) Set the **CH 1 VOLTS/DIV** control for "2 V". This will allow you to see the reverse part of the curve. Turn on the power again. Describe your observation.

Conclusion for Part 2

Questions for Part 2

1) Explain why the ground is between the components in this experiment and not on the source.

2) Could the circuit in this experiment be used to plot the characteristic curve for a resistor? Explain your answer.

3) If the 1.0 kΩ resistor is replaced with a 5.0 kΩ resistor and the **CH 1 VOLTS/DIV** control is set to 1 V/div, what is the effective **CURRENT/DIV** setting?

Name _____   Class _____

Date _____   Instructor_____

Devices Experiment 2 - Diode Applications

2.1   Introduction

In analyzing diode circuits, it is useful to use models of behavior as discussed in Floyd's text. This experiment assumes that a forward-biased diode will drop 0.7 V in all circuits. Other effects can be ignored for analysis work in most cases.

In Part 1 of this experiment, you will observe and test several Multisim diode rectifier circuits that are used to convert ac to dc. These circuits are particularly important because most electronic systems require dc. You will investigate several power supply rectifiers here and determine the line and load regulation.

In Part 2, you will focus on diode **clipping** and **clamping** circuits. Clipping circuits (also called **limiting circuits**) prevent a waveform from exceeding some particular limit, either negative or positive. Clamping circuits shift the dc level of a waveform in signal processing, wave shaping, and communication circuits.

2.2   Reading

Floyd, *Electronic Devices, 8th Edition*, Chapter 2

2.3   Key Objectives

Part 1:   Calculate and measure parameters for half-wave, full-wave, and bridge rectifier circuits including ripple voltage with capacitive-input filters.

Part 2:   Measure the effect on the output waveform with various changes to biased clipping and clamping circuits.

2.4   Multisim Files

Part 1:   *Devices_Exp_02_Part_01a*, *Devices_Exp_02_Part_01b*, and *Devices_Exp_02_Part_01c*

Part 2:   *Devices_Exp_02_Part_02a* and *Devices_Exp_02_Part_02b*

Part 1:   Diode Rectifiers

2.5   Half-Wave Rectifier with a Capacitive Filter

1)   Open the Multisim file *Devices_Exp_02_Part_01a* , which simulates a half-wave rectifier with a capacitor input filter. Diodes will have a line on one side, indicating the cathode. The purpose of the switch is to observe the circuit with and without the capacitor. For reference, Figure 2-1 shows the circuit set up on a protoboard, to illustrate the actual wiring. The transformer and fuse are not shown.

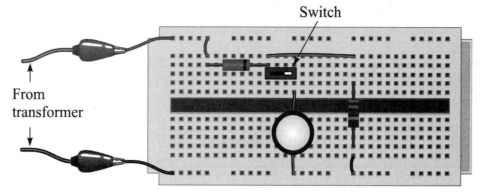

Figure 2-1: Half-Wave Rectifier Circuit on Protoboard

2) Connect the Tektronix oscilloscope so that channel 1 is across the transformer secondary (point A) and channel 2 is across the output (load) resistor (point B). SW_1 should be open.

    a. Turn on the oscilloscope.

    b. On the **TRIG** menu, verify that the trigger **Source** is **CH1**. (*Note*: Normally, if you are looking at a signal that is synchronized to the ac utility voltage, you can use "line" triggering to provide a stable trigger. Multisim does not offer this option, so we will use channel 1 for triggering and keep a reasonably large signal on it.)

    c. Activate channel 2 by pressing the **CH2 MENU** button.

    d. Set the **VOLTS/DIV** control to "5 V" for both channels.

    e. Set the **SEC/DIV** control to "5 ms".

Observations:

3) Use the spacebar to close SW_1 while observing the oscilloscope.

Observations:

4) Assume an ideal transformer for the following calculations.

    a. Calculate the peak secondary voltage. Record this value in the "Computed" column of $V_{sec(p)}$ in Table 2-1.

    b. Calculate the peak-to-peak ripple voltage across the load with SW_1 closed. The ripple voltage is given by the approximate equation

$$V_{r(pp)} \approx \left(\frac{1}{fR_LC}\right)V_{p(rect)}$$

For the half-wave circuit, the frequency is 60 Hz. Record the value in the "Computed" column of $V_{r(pp)}$ for the 1000 µF capacitor in Table 2-1. Remember to include the diode drop to account for the peak rectified voltage.

5) Using the Tektronix oscilloscope, measure the peak secondary voltage. Record the measured value of the ac in the "Measured" column for $V_{sec(p)}$ in Table 2-1.

6) With SW_1 closed, the output is a dc level with ripple. Measure the peak-to-peak ripple voltage using channel 2 as follows:

    a. On the **CH 2 MENU**, select **AC Coupling**. This blocks the dc component, allowing you to "magnify" the ripple voltage.

    b. Set the **CH2 VOLTS/DIV** control to "100 mV".

Record the measured peak-to-peak ripple voltage in the measured column of $V_{r(pp)}$ for the 1000 µF capacitor in Table 2-1.

7) Replace the 1000 µF capacitor with a 3300 µF capacitor. Calculate the expected ripple for this case. Record this value in the "Calculated" column of $V_{r(pp)}$ for the 3300 µF of Table 2-1.

8) Measure the peak-to-peak ripple voltage for the case with the 3300 µF capacitor. Record this value in the "Measured" column of $V_{r(pp)}$ for the 3300 µF in Table 2-1.

Table 2-1: Half-Wave Rectifier

$V_{sec(p)}$		$V_{r(pp)}$ (1000 µF capacitor)		$V_{r(pp)}$ (3300 µF capacitor)	
Computed	Measured	Computed	Measured	Computed	Measured

2.6 Full-Wave Rectifier with a Capacitive Filter

1) Open the Multisim file *Devices_Exp_02_Part_01b*, which simulates a full-wave rectifier with a capacitor input filter. Notice that in this configuration, the center-tap of the transformer is grounded, whereas in the half-wave circuit, one side of the transformer was grounded. This means the output voltage is one-half the value in the half-wave circuit. For reference, Figure 2-2 shows the circuit set up on a protoboard to illustrate one way of actually wiring the circuit. The transformer and fuse are not shown. There are, of course, other ways that you could connect the circuit on a protoboard.

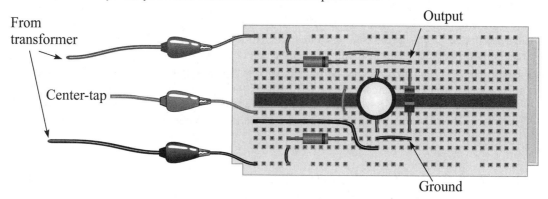

Figure 2-2: Full-Wave Rectifier Circuit on Protoboard

2) Assume an ideal transformer for the circuit.

   a. Calculate the peak secondary voltage with respect to the center tap (ground). Record this value in the "Computed" column for $V_{sec(p)}$ in Table 2-2.

   b. Calculate the peak-to-peak ripple voltage at the output for the 1000 μF capacitor and the 330 Ω load resistor. The ripple voltage is given by the same approximate equation as before, but notice that the frequency is 120 Hz and the peak rectified voltage is not the same as in the half-wave case because of the change in ground reference. You will need to account for one diode drop.

$$V_{r(pp)} \approx \left(\frac{1}{fR_LC}\right)V_{p(rect)}$$

   Enter the value in the "Computed" column for $V_{r(pp)}$ for the 1000 μF capacitor of Table 2-2.

3) Connect the Tektronix oscilloscope so that channel 1 is measuring one of the secondary leads. Connect channel 2 so that it is measuring the output.

   a. Turn on the oscilloscope.

   b. Set the **CH1 VOLTS/DIV** control to "5 V".

   c. On the **TRIG** menu, verify that the trigger **Source** is **CH1**.

   d. Set the **SEC/DIV** control to "5 ms".

   Record the measured peak ac voltage in the "Measured" column for $V_{sec(p)}$ in Table 2-2.

4) Measure the peak-to-peak ripple voltage across the load by setting up CH2 as before. Record the measured value in the "Measured" column for $V_{r(pp)}$ for the 1000 μF capacitor in Table 2-2. To make an accurate measurement, you may want to adjust the **VOLTS/DIV** control for CH2 to show a larger signal.

5) Replace the 1000 μF capacitor with a 3300 μF capacitor. Calculate the expected ripple for this case. Record the computed value in the "Calculated" column of $V_{r(pp)}$ for the 3300 μF capacitor in Table 2-2.

6) Measure the peak-to-peak ripple voltage for the case with the 3300 μF capacitor. Record the measured value in the "Measured" column of $V_{r(pp)}$ for the 3300 μF capacitor in Table 2-2.

Table 2-2: Full-Wave Rectifier

$V_{sec(p)}$		$V_{r(pp)}$ (1000 µF capacitor)		$V_{r(pp)}$ (3300 µF capacitor)	
Computed	Measured	Computed	Measured	Computed	Measured

2.7 Bridge Rectifier with a Capacitive Filter

1) Open the Multisim file *Devices_Exp_02_Part_01c*, which simulates a bridge rectifier circuit with a capacitor-input filter that can be switched in or out. Notice that in this configuration, the center-tap of the transformer is *not* connected. If it is accidentally connected to ground, a fuse will blow!

2) Connect the Tektronix oscilloscope so that channel 1 is observing the signal on one side of the transformer (point A) and channel 2 is across the output (load) resistor (point B). Notice that the voltage observed on CH1 is *not* the total secondary voltage because of the location of the ground reference. Test the effect of opening and closing SW_1 while observing the signals on the oscilloscope.

Observations:

3) Assume an ideal transformer for the circuit.

   a. Calculate the ac secondary voltage with respect to ground. Enter the computed value in the "Computed" column of $V_{sec(p)}$ in Table 2-3.

   b. Calculate the peak-to-peak ripple voltage at the output for the 1000 µF capacitor and the 330 Ω load resistor. The ripple voltage is given by the same approximate equation as before. Keep in mind, however, that there are now two diodes conducting at the same time in the bridge rectifier.

$$V_{r(pp)} \approx \left(\frac{1}{fR_L C} \right) V_{p(rect)}$$

Enter the computed value in the third column of Table 2-3.

4) Connect the Tektronix oscilloscope and measure the peak secondary voltage as before. Record this value in the "Measured" column of $V_{sec(p)}$ for the 1000 µF capacitor in Table 2-3.

5) Measure the peak-to-peak voltage ripple voltage as before. Record this value in the "Measured" column of $V_{r(pp)}$ for the 1000 µF capacitor in Table 2-3.

6) Replace the 1000 µF capacitor with a 3300 µF capacitor. Calculate the expected ripple for this case. Enter the computed value in the "Calculated" column of $V_{r(pp)}$ for the 3300 µF capacitor in Table 2-3.

7) Measure the peak-to-peak ripple voltage for the case with the 3300 µF capacitor. Record this value in the "Measured" column of $V_{r(pp)}$ for the 3300 µF cpacitor in Table 2-3.

Table 2-3: Bridge Rectifier

$V_{sec(p)}$		$V_{r(pp)}$ (1000 µF capacitor)		$V_{r(pp)}$ (3300 µF capacitor)	
Computed	Measured	Computed	Measured	Computed	Measured

Conclusion for Part 1

Questions for Part 1

1) Assume you decided to construct the circuits in this experiment. What is the minimum power rating you would need for R_L? Explain your answer.

2) What possible problem would account for higher ripple voltage in a bridge rectifier?

3) In step 1 of Section 2.7, it was stated that if the center tap is grounded, a fuse would blow. Explain.

Part 2:   Diode Clipping and Clamping Circuits

1) Open the Multisim file *Devices_Exp_02_Part_02a*. Notice that the dc source is set to 0 V. Connect the Tektronix oscilloscope so that channel 1 is connected to the ac source and channel 2 is connected to R_L.

   a. Turn on the oscilloscope.

   b. Activate channel 2 by pressing the **CH2 MENU** button.

   c. Set the **VOLTS/DIV** control to "5 V" for both channels.

   d. Set the **SEC/DIV** control to "200 μs".

   e. From the **MATH** menu, choose **Operation (−)**.

   f. Select **CH1−CH2** using the pushbutton. Your active selection will be highlighted.

   Note: Whenever you want to view the actual difference in two signals, set both channels to have the same volts per division setting.

2) There should be three signals on the display. Channel 1 shows the source voltage, channel 2 shows the load voltage, and the channel labeled **M** (for Math) is showing the difference between these two channels, which represents the voltage across R_1. You can separate the signals for easier viewing using the vertical position controls for each channel. Notice that the algebraic sum or these signals is zero, in accordance with Kirchhoff's voltage law.

3) Sketch the waveforms relative to the source voltage in Figure 2-3. Label each axis of the figure.

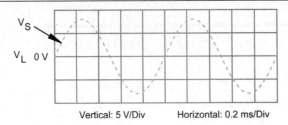

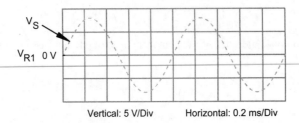

Figure 2-3: Observed Clipper Waveforms

Note: For steps 4 through 7, turn off the simulation before changing the circuit. After changing the circuit, restart the simulation.

4)   Set the dc power supply to +2.0 V. Observe and describe the same signals as before.

5)   Reverse the diode for the clipping circuit. To reverse a component, first disconnect it, select it, and then click "ALT+Y". Try different dc voltages and describe the results. To change the dc voltage, double-click the dc source and set a new value in the properties window. Describe the signals.

6)   Restore the diode to its original configuration (anode connected to the output). Then reverse the positive dc power supply so that it becomes a negative supply. Reconnect the dc supply. Again, vary the dc voltage and describe the results.

7)   Open the Multisim file *Devices_Exp_02_Part_02b*. Again, set the dc source to 0 V. Connect the channel 1 of the Tektronix scope to the ac source and channel 2 to R_L. This circuit is a clamping circuit. Set up the scope the same way as you did in step 1, using the **MATH** menu to view **CH1−CH2**. This time the difference signal is the voltage on the capacitor. Describe the three signals that you see. You can measure the voltage across the capacitor with a dc multimeter to confirm your scope reading.

8)   Turn off power to the circuit, change the dc voltage to +0.7 V, and reactivate the circuit. Describe what happens.

9) Reduce the ac signal to 5.0 V_p. Describe what happens to the clamping level and the dc voltage on the capacitor.

Conclusion for Part 2

Questions for Part 2

1) How can you modify the basic circuit in this experiment to form a positive clipping circuit with a clipping level at 0 V?

2) From the data you graphed in Figure 2-3, show that Kirchhoff's voltage law is satisfied when the input is at its peak positive voltage and its peak negative voltage.

3) If you wanted to change the clamping circuit in step 7 from a positive clamping circuit to a negative clamping circuit with a clamping level of 0 V, you need to reverse the diode and the capacitor (if you build an actual circuit.)

   a. Why is it necessary to reverse the capacitor?

   b. What should you do to the dc power supply to complete the modification?

Devices Experiment 3 - Bipolar Junction Transistors

3.1   Introduction

A bipolar junction transistor (BJT) is a three-terminal device capable of amplifying an ac signal. BJTs are current amplifiers composed of three layers of alternating *n* and *p* material forming a "sandwich" with a thin base layer in the center. A small base current is amplified to a larger current in the collector-emitter circuit. An important BJT characteristic is the dc current gain, called the dc beta (β_{DC}), which is the ratio of collector current to base current. Another parameter is the ac beta (also called small-signal current gain), which is a *change* in collector current divided by a *change* in base current. In Part 1 of this experiment, you will plot the characteristic curve for a BJT and use the plot to determine both the β_{DC} and the β_{ac}. The characteristic curve is a plot of the collector-emitter voltage as a function of the collector current for a constant base current. In Part 2, you will learn how to plot the curves automatically on an oscilloscope, simulating an instrument that is called a curve tracer (or the **IV-Analysis** tool in Multisim). Commercial curve tracers can display a family of curves simultaneously. The simulation will show only one at a time, but the concept of how a curve tracer works should be clear.

For a transistor to amplify, dc voltages are required. These dc voltages are referred to as bias voltages. In Part 1, the dc voltages are supplied by two separate supplies, allowing you to have control over the input and output current. In Devices Experiment 4, you will see more useful ways of supplying the bias voltages in actual amplifiers.

3.2   Reading

Floyd, *Electronic Devices, 8th Edition*, Chapter 4

3.3   Key Objectives

Part 1:   Measure the collector characteristic curves for a BJT and compare the dc and ac beta obtained from the curves with the manufacturer's specification sheet.

Part 2:   Connect a circuit that will plot the collector characteristic curves for both an *npn* and a *pnp* transistor on an oscilloscope.

3.4   Multisim Files

Part 1:   *Devices_Exp_03_Part_01*

Part 2:   *Devices_Exp_03_Part_02*

Part 1:   BJT Characteristic Curve (Manual)

1)   Open the Multisim file *Devices_Exp_03_Part_01*, which shows a test setup for a bipolar junction transistor (BJT). For reference, Figure 3-1 shows the same setup on a protoboard to illustrate the actual wiring. Power supply voltages are given as V_1 (base supply) and V_2 (collector supply).

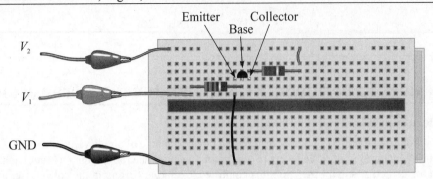

Figure 3-1: BJT Circuit on Protoboard

2) The circuit is a test setup to measure the characteristic curve for a 2N3904 *npn* transistor. The necessary meters have been connected to simplify the setup.

    a. Set V_1 for 2.38 V and V_2 for 5 V.

    b. Read the voltage on XMM_1 and use Ohm's law to calculate the base current. I_B. You should be able to verify that it is 50 µA.

    c. Record to three significant digits the collector-emitter voltage indicated on XMM_2 and the collector current on ammeter U_1 in the "$I_B = 50$ µA" column in Table 3-1.

3) Turn off power to the circuit. Keep V_1 set to 2.38 V but change V_2 to the next value in Table 3-1 and restore power. Record to three significant digits the collector-emitter voltage indicated on XMM_2 and the collector current on ammeter U_1 in Table 3-1. Repeat for each value of V_2 listed in Table 3-1.

4) Adjust V_1 for 4.08 V. Read the voltage on XMM_1 and use Ohm's law to calculate the base current, I_B. You should be able to verify that it is now 100 µA.

    a. Set V_2 to 5.0 V.

    b. Notice that the reading on the ammeter is now rounded to two significant figures (because it is greater than 9.99 mA), which is not as precise as it should be for determining the points on the curve. To obtain three significant figures, subtract the reading on XMM_2 from V_2 and divide by the value of R_C (this is just Ohm's law applied to R_C). Record to three significant digits the voltage and current in the "$I_B = 100$ µA" column in Table 3-1.

    c. Repeat this for the other values of V_2 listed in Table 3-1 for $I_B = 100$ µA.

5) Adjust V_1 for 5.76 V. Read the voltage on XMM_1 and use Ohm's law to calculate the base current, I_B. You should be able to verify that it is 150 µA.

    a. Set V_2 to 5.0 V.

    b. Apply Ohm's law to R_C and record to three significant digits the collector-emitter voltage and collector current in the "Base Current = 150 µA" column in Table 3-1.

    c. Repeat this for the other values of V_2 listed in Table 3-1 for $I_B = 150$ µA.

Table 3-1: Measured Characteristic Curve Data

V_2	$I_B = 50$ µA ($V_1 = 2.38$ V)		$I_B = 100$ µA ($V_1 = 4.08$ V)		$I_B = 150$ µA ($V_1 = 5.76$ V)	
	V_{CE}	I_C	V_{CE}	I_C	V_{CE}	I_C
5.0 V						
7.5 V						
10 V						
15 V						
20 V						

6) Plot the data from Table 3-1 in Plot 3-1 (I_C versus V_{CE}). You will have three lines representing the characteristic curve for a given base current on the same graph. Label each line for the base current and add a label for the overall plot.

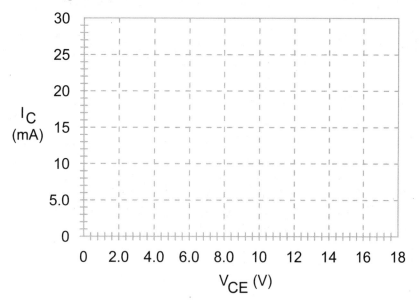

Plot 3-1: I_C **vs.** V_{CE} **Characteristic Curves**

Conclusion for Part 1

Questions for Part 1

1) Does the experimental data indicate that β_{DC} is a constant at all points? Does this have any effect on the linearity of the transistor?

2) From your data, what is the approximate β_{DC} for the transistor? Use the curve for $I_B = 100\ \mu A$ at $V_{CE} = 8$ V to estimate your answer.

3) What collector current would you expect if $V_{CE} = 10$ V and $I_B = 75\ \mu A$?

Part 2: BJT Characteristic Curve on the Oscilloscope

In this part, you will set up a scope in a circuit in Multisim to view the characteristic curve for a 2N3904 *npn* transistor on the scope. Then you will modify the circuit to display a *pnp* transistor curve.

1) Open the Multisim file *Devices_Exp_03_Part_02*. Connect the Tektronix oscilloscope so that channel 1 is connected to the collector of the transistor and channel 2 is connected to the top of R_C.

a. Turn on the oscilloscope.

b. In the **DISPLAY** menu, choose **Format XY**.

c. In the **CH 1 MENU** choose **Coupling Ground**. This disconnects the *x*-input and allows you to see where ground is positioned on the *x*-axis.

d. Activate the circuit. You should see a vertical line near the center of the screen. Move the **CH 1 POSITION** control until the line is positioned on the left side of the screen in line with the first vertical line on the display. This will cause all divisions on the *x*-axis to have positive values starting from the left side.

e. In the **CH 1 MENU** choose **Coupling DC**. (This reconnects the channel to the input.)

f. Set the **CH1 VOLTS / DIV** to "2 V".

g. In the **CH 2 MENU** choose **Coupling Ground**. This disconnects the *y*-input enabling you to see where ground is positioned on the *y*-axis.

h. Move the **CH 2 POSITION** control until the line is positioned along the bottom of the screen. This will cause all divisions on the *y*-axis to have positive values.

i. In the **CH 2 MENU** choose **Coupling DC** and **Invert On**. This causes the scope to display the increase in current in the positive direction.

j. You should observe one of the characteristic curves for the transistor. You can change the base current by selecting different switch positions on the input. The multimeter allows you to check the exact base current that is selected.

k. Set the **CH2 VOLTS/DIV** to "2 V".

2) Sketch the three characteristic curves on Figure 3-2, based on the scope display. Label the axes and identify each curve. The Multisim display for the $I_B = 50\ \mu A$ curve is shown below for reference.

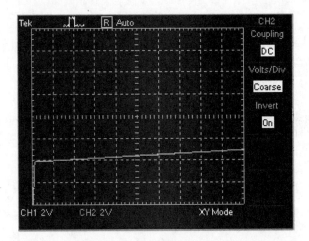

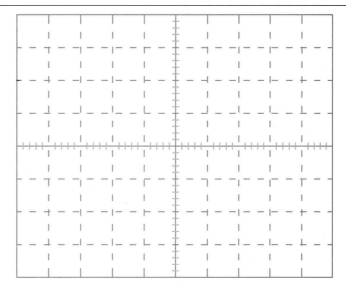

Figure 3-2: NPN I_C vs. V_{CE} Characteristic Curves

3) Turn off the circuit. Replace the 2N3904 *npn* transistor with a 2N3906 *pnp* transistor. Decide what other changes are necessary to the circuit to be able to show the characteristic curves for the *pnp* transistor. Describe the changes you made to the circuit:

4) To view the *pnp* characteristics, you will need to modify the scope settings. List the changes made to the scope settings. (Note that the volts per division controls can be left at "2 V", but a small part of the largest curve will be cut off.)

5) Sketch the three characteristic curves for the *pnp* transistor on Figure 3-3, based on the scope display. Label the axes and identify each curve. The Multisim display for the $I_B = 50$ µA curve is shown below for reference.

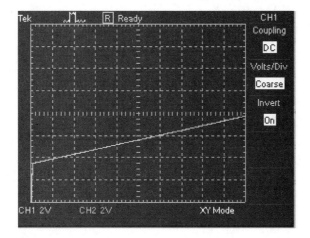

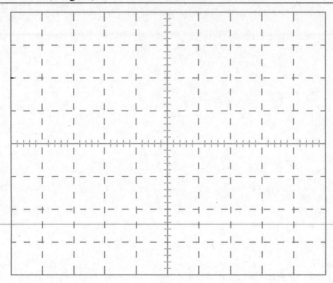

Figure 3-3: PNP I_C vs. V_{CE} Characteristic Curves

Conclusion for Part 2

Questions for Part 2

1) How did the shape of the characteristic curve differ between the *npn* transistor and the *pnp* transistor?

2) From the curves you plotted in this part of the experiment, compare the β_{DC} of the *npn* and the *pnp* transistors at a base current of 100 μA and a V_{CE} of 10 V.

3) The transconductance, g_m, of a BJT transistor is defined as the change in collector current, ΔI_C, divided by the change in collector-emitter voltage, ΔV_{CE}, so that $g_m = \Delta I_C / \Delta V_{CE}$. Use your curves in Figure 3-2 to calculate the transconductances for the *npn* transistor for $I_B = 50$ µA, $I_B = 100$ µA, and $I_B = 150$ µA.

4) Use your curves in Figure 3-3 to calculate the transconductances for the *pnp* transistor for $I_B = 50$ µA, $I_B = 100$ µA, and $I_B = 150$ µA.

5) How do the calculated transconductance values compare for the *npn* and *pnp* transistors?

Devices Experiment 4 - BJT Biasing

4.1   Introduction

For a transistor to operate in the linear region, it is necessary to forward-bias the base-emitter junction and to reverse-bias the base-collector junction. The purpose of bias is to provide dc voltages to set up the quiescent conditions that establish the operating point on the characteristic curve for the device.

Four common bias circuits used for establishing the necessary conditions for linear operation will be investigated in Part 1 of this experiment. The circuits are (1) voltage-divider bias, (2) base bias, (3) emitter bias and (4) collector-feedback bias. In actual lab work, you will find variations in measured voltages and currents for any given circuit due to manufacturing differences in transistors, even when the same type of transistor is used. To provide for these normal manufacturing differences, two different transistors are used in this experiment – one is the 2N3903 *npn* transistor and the second one is the 2N3904 *npn* transistor Although both of these transistors are general-purpose amplifiers, the 2N3904 has β_{DC} that is nearly twice that of the 2N3903. This will illustrate differences that you will encounter with similar transistors with different β_{DC}.

In Part 2, a transistor switching circuit is investigated. The bias circuits for switching can be even simpler than those for linear operation because the transistor is ideally operated in either *cutoff* (no conduction) or *saturation* (maximum conduction). The first circuit tested is not ideal because it can operate *between* cutoff and saturation, a condition not desirable in a switching circuit. After testing it, you will see how to make improvements to this basic circuit to avoid operating in the linear region. The switching points are investigated and the concept of hysteresis is introduced, where two switching points are used, depending on whether the circuit is turning on or off.

4.2   Reading

Floyd, *Electronic Devices, 8th Ed.*, Chapter 5

4.3   Key Objectives

Part 1:   Analyze four types of transistor bias circuits: voltage-divider bias, base bias, emitter bias, and collector-feedback bias.

Part 2:   Analyze a BJT transistor switching circuit for switching thresholds including one with hysteresis.

4.4   Multisim Files

Part 1:   *Devices_Exp_04_Part_01a*, *Devices_Exp_04_Part_01b*, *Devices_Exp_04_Part_01c*, and *Devices_Exp_04_Part_01d*

Part 2:   *Devices_Exp_04_Part_02a*, *Devices_Exp_04_Part_02b*, and *Devices_Exp_04_Part_02c*

Part 1:   Bias Methods

4.5   Voltage-Divider Bias

1) Open the Multisim file *Devices Exp_04_Part_01a*. The circuits in this experiment all have two transistors as explained in the Introduction. The first file has two different *npn* transistors set up with identical voltage-divider bias circuits.

2) Use the following steps to calculate all the values in the "DC Parameters" column in Table 4-1. Record to three significant digits your values in the "Computed Value" column of Table 4-1.

   a.   Apply the voltage-divider rule to calculate the base voltage, V_B. In this case, an unloaded divider can be assumed. (The actual voltage will be a little lower due to loading.)

   b.   Subtract 0.7 V to calculate the emitter voltage, V_E.

   c.   Apply Ohm's law to find the emitter current, I_E. Assume it is the same as I_C.

    d.   Apply Ohm's law to find the voltage across the collector resistor, V_{RC}.

    e.   Subtract V_{RC} from V_{CC} to obtain V_C.

3)   Turn on power to the circuit. Use a DMM to measure the voltages listed for both Q_1 and Q_2. Record to three significant digits your readings in Table 4-1. In practical work, it is not usually necessary to measure current as it involves opening the circuit to insert the ammeter. Because it is simple to do in Multisim, the collector circuit includes an ammeter, so that you can measure I_C directly.

Table 4-1: Voltage-Divider Bias

DC Parameter	Computed Value	Measured Value	
		Q_1 (2N3903)	Q_2 (2N3904)
V_B			
V_E			
$I_E \approx I_C$			
V_{RC}			
V_C			

4.6   Base Bias

1)   Open the Multisim file *Devices_Exp_04_Part_01b*. In this file, the two circuits use base bias; notice that the base resistor is much larger than the two bias resistors used in voltage-divider bias. While this offers the advantage of higher input resistance, it is more than offset by shifts in operating point due to β_{DC} differences. Calculate all the values in the "DC Parameter" column of Table 4-2. Record to three significant digits your values in the "Computed Value" column of Table 4-2.

    a.   Calculate the collector current from the equation:

$$I_C \approx I_E = \frac{V_{CC} - V_{BE}}{\dfrac{R_B}{\beta_{DC}} + R_E}$$

    You can derive this equation by applying Kirchhoff's voltage law through the base-emitter circuit. For the calculated value, assume that $\beta_{DC} = 100$.

    b.   Calculate V_E by applying Ohm's law to R_E.

    c.   Apply Ohm's law to find the voltage across the collector resistor, V_{RC}.

    d.   Calculate V_C by subtracting V_{RC} from V_{CC}.

    e.   Calculate V_{CE} by subtracting V_E from V_C.

2)   Turn on power to the circuit. Use a DMM to measure the voltages listed for both Q_1 and Q_2 and read the collector current on the ammeter. Record to three significant digits your readings in Table 4-2.

Table 4-2: Base Bias

DC Parameter	Computed Value	Measured Value	
		Q_1 (2N3903)	Q_2 (2N3904)
$I_E \approx I_C$			
V_E			
V_{RC}			
V_C			
V_{CE}			

4.7   Emitter Bias

1) Open the Multisim file *Devices_Exp_04_Part_01c*. In this file, the two circuits use emitter bias; this form of bias is very stable, but the drawback is that it generally requires two power supplies (an exception is for a specific type of amplifier called an emitter-follower or common-collector amplifier, introduced in the next experiment). Calculate all the values listed in the "DC Parameter" column of Table 4-3. Record to three significant digits your values in the "Computed Value" column of Table 4-3.

   a.   Calculate the collector current from the equation:

$$I_C \approx I_E = \frac{-V_{CC} - V_{BE}}{\dfrac{R_B}{\beta_{DC}} + R_E}$$

   You can derive this equation by applying Kirchhoff's voltage law to the base-emitter circuit. For the calculated value, assume $\beta_{DC} = 100$.

   b.   Calculate V_{RE} by applying Ohm's law to R_E (and assuming that $I_E \approx I_C$).

   c.   Calculate V_E by (algebraically) adding V_{RE} to V_{EE}.

   d.   Calculate V_{RC} by applying Ohm's law to R_C.

   e.   Calculate V_C by subtracting V_{RC} from V_{CC}.

   f.   Calculate V_{CE} by subtracting V_E from V_C.

2) Turn on power to the circuit. Use a DMM to measure the voltages listed for both Q_1 and Q_2 and read the collector current on the ammeter. Record your readings to three significant digits in Table 4-3.

Table 4-3: Emitter Bias

DC Parameter	Computed Value	Measured Value	
		Q_1 (2N3903)	Q_2 (2N3904)
$I_E \approx I_C$			
V_{RE}			
V_E			
V_{RC}			
V_C			
V_{CE}			

4.8   Collector-Feedback Bias

1) Open the Multisim file *Devices_Exp_04_Part_01d*. In this file, the two circuits use collector-feedback bias; this form of bias is more stable than base bias and uses the same number of components. The Multisim circuits have an emitter resistor that is optional, but is used for gain stability. Calculate all the values listed in the "DC Parameter" column of Table 4-4. Record to three significant digits your values in the "Computed Value" column of Table 4-4.

   a.   Calculate the collector current from the equation:

$$I_C \approx I_E = \frac{V_{CC} - V_{BE}}{R_C + \dfrac{R_B}{\beta_{DC}} + R_E}$$

   You can derive this equation by applying Kirchhoff's voltage law through the collector feedback resistor and the base-emitter junction. For the calculated value, assume that $\beta_{DC} = 100$.

   b.   Calculate V_{RE} by applying Ohm's law to R_E (and assuming $I_E \approx I_C$).

   c.   Calculate V_E by (algebraically) adding V_{RE} to V_{EE}.

    d.   Calculate V_{RC} by applying Ohm's law to R_C.

    e.   Calculate V_C by subtracting V_{RC} from V_{CC}.

    f.   Calculate V_{CE} by subtracting V_E from V_C.

2)   Turn on power to the circuit. Use a DMM to measure the voltages listed for both Q_1 and Q_2 and read the collector current on the ammeter. Record to three significant digits your readings in Table 4-4.

Table 4-4: Collector-Feedback Bias

DC Parameter	Computed Value	Measured Value	
		Q_1 (2N3903)	Q_2 (2N3904)
$I_E \approx I_C$			
V_{RE}			
V_E			
V_{RC}			
V_C			
V_{CE}			

Conclusion for Part 1

Questions for Part 1

1)   From the data you took, which biasing method has the lowest dependency on β_{DC}?

2)   What change would you make if you wanted even less dependency on β_{DC} for voltage-divider bias?

3)   What is the disadvantage of using smaller divider resistors for voltage-divider bias?

Part 2:   Transistor Switching Circuits

4.9   Basic Switching Circuit

1)   Open the Multisim file *Devices Exp_04_Part_02a*. The circuit is a basic transistor switching circuit using voltage-divider bias that can be adjusted by varying R_2. As a switching circuit, it does not have ideal characteristics as explained in the Introduction. Ideally it should be either *on* or *off*. To simplify monitoring, a voltmeter is connected to the base circuit and an ammeter is connected in the collector circuit.

2)   Calculate the parameters listed in Table 4-5 at cutoff and saturation as follows:

    a.   Assume that V_{LED} for both cutoff and saturation is 1.7 V. (This is the approximate forward bias drop.) At cutoff, the remaining voltage that is supplied by V_{CC} is across the transistor. Calculate $V_{CE(cutoff)}$ by applying Kirchhoff's voltage law to the collector circuit.

    b.   At saturation, the transistor is fully on. For this experiment, assume that $V_{CE(sat)} = 0.1$ V (a typical value).

    c.   Calculate $V_{RC(sat)}$ by applying Kirchhoff's voltage law to the collector circuit with $V_{CE(sat)}$ equal to 0.1 V and $V_{LED} = 1.7$ V.

    d.   Calculate I_{sat} by applying Ohm's law to $V_{RC(sat)}$.

3)   Connect a DMM between the collector and the emitter of the transistor. Turn on power to the circuit. When power is applied, R_2 will be set by Multisim to its default value of 50%. If you hover over R_2, a slider bar appears and you can vary R_2 to any desired percentage. Start at the 50% default resistance, which will cause the transistor to be biased off because the voltage is too small to forward-bias the base-emitter junction. The DMM will show the measured value of $V_{CE(cutoff)}$. Record to three significant digits the reading of $V_{CE(cutoff)}$ in Table 4-5.

4)   *Slowly* increase the percentage of resistance set on R_2 using the "A" key to increase the percentage by 1% each time. (Note that "SHIFT + A" decreases the percentage.) You need to do this slowly to allow Multisim time to calculate new conditions as you change the potentiometer. Observe the collector current as you gradually increase the setting. You will also see the LED comes on at some point. In the "real world", this is more gradual than in the simulation; Multisim has a specific threshold and cannot show gradual brightening but the collector current is a good indication of what is happening. Summarize your observations.

5)   Increase R_2 to the maximum level (100%). Record the value of $V_{CE(sat)}$ Table 4-5.

6)   Turn off the power to the circuit and connect the DMM across R_C. Turn on power and measure $V_{RC(sat)}$. Apply Ohm's law to R_C to obtain a more accurate reading of I_{sat} than the ammeter provides. Record the readings in Table 4-5.

Table 4-5: Basic Switching Circuit

Parameters	Computed Value	Measured Value
$V_{CE(cutoff)}$		
$V_{CE(sat)}$		
$V_{RC(sat)}$		
I_{sat}		

4.10 Improved Switching Circuit

You may have noticed that the collector current changed over a range of values as you varied R_2. A simple change to this circuit can make the switching point much more abrupt. The change is to add a second transistor that is biased from the collector current of the first transistor (this is called *direct coupling*).

1)   Open the Multisim file *Devices_Exp_04_Part_02b*. The response will be different than for the first circuit, and the LED has been moved to the collector circuit of Q_2.

2)   As before, start R_2 at 50% and gradually increase the percentage resistance using the "A" key, allowing Multisim to settle down after each adjustment. Note how the circuit responds to a small change in the input. Summarize your observations.

4.11 Switching Circuit with Hysteresis

In the last circuit, the switching threshold was distinct; however, the threshold voltage is low and it is susceptible to noise or small variations on the input, which can cause the circuit to change states if the input remains near the threshold. The circuit in this section has a modified bias scheme for Q_1 and a common-emitter resistor, R_E, that will raise the threshold voltage. Because of the different saturation currents of the two transistors, the threshold will be different when the output is in cutoff than when the output is saturated. This is a very useful feature called *hysteresis*, which has the effect of providing two different switching thresholds.

1) Open the Multisim file *Devices_Exp_04_Part_02c*. In addition to the emitter resistor, the potentiometer is larger to provide a greater range for the input. This time, keep watching the input voltage as you slowly raise the voltage using the "A" key. Just at the point that the LED goes out, the input voltage will drop, because the collector (and hence the emitter) current changes, which affects the input loading. Determine V_{IN} *just before* the LED goes out, and record this as "V_{IN} (upper threshold)" in Table 4-6. (You may need to repeat this to get the correct reading.)

2) After the LED goes out, lower the input voltage using the "Shift + A" keys (allowing time for Multisim to settle.) Note the voltage *just before* the LED turns on. Record this voltage as "V_{IN} (lower threshold)" in Table 4-6.

Table 4-6: Switching Circuit with Hysteresis

Quantity	Measured Value
V_{IN} (upper threshold)	
V_{IN} (lower threshold)	

Conclusion for Part 2

Questions for Part 2

1) In the improved switching circuit, two transistors were used.

    a. What type of bias is used by Q_1?

    b. What type of bias is used by Q_2?

2) For the circuit with hysteresis, what do you expect to happen if R_E is larger?

Devices Experiment 5 - BJT Amplifiers

5.1   Introduction

The ac analysis of BJT circuits depends on whether the circuit is a small-signal or large-signal circuit and the frequency range. High-frequency circuits are analyzed differently than low-frequency circuits. The circuits in this experiment can use low-frequency techniques, which can be analyzed with either the r_e model (referred to as the r-parameters) or the hybrid model (referred to as h-parameters). This experiment focuses on the r-parameter model, which is simple to use.

In Part 1, you will calculate the dc and ac parameters for a common-emitter (CE) amplifier, and then test the amplifier using Multisim.

In Part 2, a common-collector (CC) amplifier is investigated using a *pnp* transistor. The equations for the *npn* transistor circuit can be applied to the *pnp* except keep in mind that the emitter voltage is 0.7 V *larger* than the base voltage because of the *pnp* transistor. The CC amplifier is then combined with the CE amplifier, forming a two-stage direct-coupled amplifier, which is investigated. Direct coupling results in an overall simpler circuit because the second stage does not require a bias network or coupling capacitor. Designing any circuit involves design trade-offs, but the final result of the two-stage amplifier is an improved circuit that takes advantage of the best characteristics of both the CE and CC amplifiers.

5.2   Reading

Floyd, *Electronic Devices*, *8th Edition*, Chapter 6

5.3   Key Objectives

Part 1:   Analyze the dc and ac parameters for a common-emitter (CE) amplifier and compare your calculated values to the simulation in Multisim.

Part 2:   Analyze the dc and ac parameters for a common-collector (CC) amplifier and compare your calculated values to the simulation in Multisim. Test a two-stage amplifier using the CE and CC amplifier in a direct-coupled circuit.

5.4   Multisim Files

Part 1:   *Devices_Exp_05_Part_01*

Part 2:   *Devices_Exp_05_Part_02a* and *Devices_Exp_05_Part_02b*

Part 1:   The Common-Emitter Amplifier

5.5   Calculations and Measurements

1)   Open the Multisim file *Devices_Exp_05_Part_01*. For reference, Figure 5-1 shows the circuit on a protoboard. Calculate the dc quantities listed in Table 5-1 for the CE amplifier. Record your calculations in Table 5-1. The steps for each quantity are listed:

   a.   Calculate V_B using the voltage-divider rule. Because $\beta R_E \geq 10R_2$, you can assume an unloaded voltage divider.

   b.   Subtract V_{BE} from V_B to obtain V_E.

   c.   Apply Ohm's law to the emitter resistors to find I_E. Assume that it is the same as I_C.

   d.   Calculate V_{RC} by applying Ohm's law to R_C.

   e.   Calculate V_C by subtracting V_{RC} from V_{CC}.

   f.   Calculate V_{CE} by subtracting V_E from V_C.

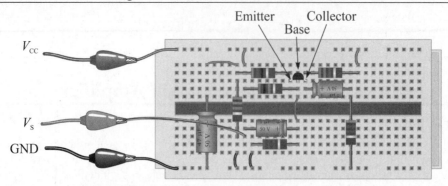

Figure 5-1: Common-Emitter Amplifier on Protoboard

2) Turn on power to the circuit. Use a DMM to measure the dc voltages listed in Table 5-1 and read the ammeter to measure the collector current. It is a good idea to turn ac signal sources to zero when you measure dc quantities, because some circuits can convert part of the ac signal to dc and affect your results. Record to three significant digits your readings in Table 5-1. It is not always practical to measure current, as this involves opening the circuit to insert the ammeter. Because it is simple to do in Multisim, the collector circuit includes an ammeter, so that you can measure I_C directly.

Table 5-1: Common Emitter Amplifier DC Values

DC Parameter	Computed Value	Measured Value
V_B		
V_E		
$I_E \approx I_C$		
V_{RC}		
V_C		
V_{CE}		

3) Use the following steps to calculate all the values listed in the "AC Parameter" column of Table 5-2:

   a. The input signal, V_{in}, is already shown as 150 mV$_p$ (300 mV$_{pp}$). This is both V_{in} and the ac base voltage, V_b, because the source resistance is assumed to be zero in Multisim.

   b. Calculate r_e' from $r_e' = 25$ mV $/ I_E$. Use your calculated value of I_E for this.

   c. Calculate the ac gain from $A_v = (R_C \| R_L) / (r_e' + R_{E1})$.

   d. Multiply V_{in} by the computed voltage gain to calculate the ac voltage at the collector; this is both V_c and V_{out}.

   e. Calculate $R_{in(tot)}$ from $R_{in(tot)} = R_1 \| R_2 \| \beta_{ac}(r_e' = R_{E1})$. Assume that $\beta_{ac} = 100$.

4) Measure all the values in the "AC Parameter" column in Table 5-2 (except for r_e'). The voltage gain is measured by dividing V_{out} by V_{in}. The value of $R_{in(tot)}$ is measured indirectly as follows (using the test resistor, which was put in the circuit for that purpose).

   a. Note the amplitude of the output signal with $R_{test} = 0$ Ω. The output signal is monitored to assure that there is no clipping, which could affect the measurement.

   b. Change R_{test} to the computed value of $R_{in(tot)}$. If the output drops by exactly one-half, the computed and measured values are equal.

c. If the output is more than one-half its former value, increase R_{test}. If the output is less than one-half its former value, decrease R_{test}. Adjust R_{test} until the output drops by one-half the original value. Note the resistance of R_{test} and record it as the measured value of $R_{in(tot)}$.

Table 5-2: Common-Emitter Amplifier AC Values

AC Parameter	Computed Value	Measured Value
$V_{in} = V_b$	150 mV$_p$	
r_e'		
A_v		
$V_{out} = V_c$		
$R_{in(tot)}$		

5.6   Troubleshooting

1) Remove one end of the bypass capacitor, C_2, from the circuit, simulating an open capacitor. Measure the ac signal voltage at the transistor's base, emitter, and collector. Measure the voltage gain of the amplifier. What conclusion can you make about the amplifier's performance with C_2 open?

2) Reconnect C_2 and reduce R_L to 1.0 kΩ, simulating a change in load conditions. Observe the ac signal voltage at the transistor's base, emitter, and collector. Measure the voltage gain of the amplifier. What conclusion can you make about the amplifier's performance with R_L reduced to 1.0 kΩ?

3) Replace R_L with the original 8.2 kΩ resistor and open R_{E1}. (You can open a component using the **Fault** tab when you double-click the component.) Measure the dc voltages at the base, emitter, and collector. Is the transistor in cutoff or in saturation? Explain.

4) Replace R_{E1} and open R_2. Measure the dc voltages at the base, emitter and collector. Is the transistor in cutoff or saturation? Explain.

Conclusion for Part 1

Questions for Part 1

1) To measure the input resistance of the amplifier, you monitored the output voltage. Why is this procedure better than monitoring the base voltage?

2) What is the purpose of the unbypassed emitter resistor R_{E1}? What design advantage does it offer?

3) When the bypass capacitor, C_2, is open, you found that the gain is affected. Explain why.

Part 2:   Adding a Common-Collector Stage

5.7   CC Amplifier Calculations and Measurements

1) Open the Multisim file *Devices Exp_05_Part_02a*. The circuit is a CC amplifier using a *pnp* transistor. You will first test this circuit and then combine it with the CE amplifier from Part 1. Calculate the dc quantities listed in Table 5-3 for the CC amplifier. Record your calculations in the table. This part of the experiment illustrates a more exact method of solving the dc quantities for this circuit to give you practice in this method. To apply the more exact solution, it is simpler to find the emitter current, I_E, first. Record to three significant digits all the values in the "DC Parameter" column of Table 5-3.

2) The method described in this step starts by calculating a Thevenin circuit for the input. The following steps will guide you in finding a very close result for I_E:

   a. Mentally break the circuit at the base and look back toward the source. The Thevenin circuit will have $R_1 \| R_2$, which is the Thevenin resistance, R_{TH}.

   b. The Thevenin voltage is simply the unloaded base voltage: $V_{TH} = V_{CC} [R_2 / (R_1 + R_2)]$.

   c. Next write Kirchhoff's voltage law around the loop consisting of the Thevenin circuit and the base-emitter circuit (being careful of polarities).

   $$-V_{TH} - I_B R_{TH} - V_{BE} - I_E R_{E3} + V_{CC} = 0$$

   d. From the result of Kirchhoff's voltage law, solve for I_E.

   $$I_E = (V_{CC} - V_{TH} - V_{BE}) / [R_{E3} + (R_{TH} / \beta_{DC})]$$

   e. Be sure that you understand how the equation for I_E was derived. Then apply the equation to calculate I_E and enter the result in Table 5-3. Assume that $\beta_{DC} = 100$.

   f. Calculate V_{RE3} across R_{E3} using Ohm's law.

   g. Calculate V_E by subtracting V_{RE3} from V_{CC}.

   h. Calculate V_B by subtracting V_{BE} from V_E. (Note than V_B is less than V_E because of the *pnp* transistor.)

Table 5-3: Common Collector Amplifier DC Values

DC Parameter	Computed Value	Measured Value
$I_E \approx I_C$		
V_{RE3}		
V_E		
V_B		

3) Turn on power to the circuit. Determine I_E indirectly by measuring V_{RE} and applying Ohm's law. Use the ammeter reading to confirm this value (although the reading has fewer significant digits). The ac signal source is preset to zero so that you can measure dc quantities correctly, because the transistor can convert a portion of the ac signal to dc. Record your readings using three significant digits in the "Measured Value"

column of Table 5-3. To obtain an accurate value of the measured I_E, divide the measured voltage V_{RE3} by R_{E3}.

4) Calculate and record to three significant digits the values in the "AC Parameter" column in Table 5-4. Steps for the calculations are as follows:

    a. Assume the input voltage is set to 5.0 V_p. Table 5-4 already shows this value.

    b. Calculate r_e' from $r_e' = 25$ mV / I_E. Use your calculated value of I_E for this.

    c. Calculate the ac gain from $A_v = (R_{E3} \| R_{L2}) / (r_e' + R_{E3} \| R_{L2})$. This is equivalent to $A_v = R_e / (r_e' + R_e)$.

    d. Multiply V_{in} by the computed voltage gain to calculate V_{out}.

    e. Calculate $R_{in(tot)}$ from $R_{in(tot)} = R_1 \| R_2 \| [\beta_{ac} (r_e' + R_{E3} \| R_{L2})]$. Assume that $\beta_{ac} = 100$.

5) Measure the ac parameters in Table 5-4 (except for r_e'). You must first measure V_{out} and use the measurement to calculate A_v. The value of $R_{in(tot)}$ is measured indirectly by the method described in Part 1.

Table 5-4: Common-Collector Amplifier AC Values

AC Parameter	Computed Value	Measured Value
$V_{in} = V_b$	5.0 V_p	
r_e'		
A_v		
V_{out}		
$R_{in(tot)}$		

5.8 Cascading the CE and CC Stages

Open the Multisim file *Devices_Exp_05_Part_02b*. This circuit cascades, or "marries" the CE and CC amplifiers to create a simple two-stage amplifier. An advantage of this circuit is the elimination of the load resistor for Q_1 and the bias resistors and coupling capacitor for Q_2. The second stage has been reduced to only three components plus the transistor. There are several design considerations that go into both the CE and CC amplifiers before they can be combined effectively, including matching the bias requirement and input resistance (minus the bias resistors) of the CC amplifier to that of the CE amplifier.

1) Calculate the three significant digits the values in the "AC Parameter" column in Table 5-5 as follows:

    a. The input signal, V_{in}, is already shown as 150 mV$_p$ (300 mV$_{pp}$).

    b. Calculate the total voltage gain for the two-stage amplifier from $A_{v(tot)} = A_{v(CE)} A_{v(CC)}$. Use the calculated gain for the CE amplifier from Table 5-2 and the calculated gain for the CC amplifier from Table 5-4. This is reasonable because the input resistance of the CC amplifier (less bias resistors) is close to the original load resistance of the CE amplifier.

    c. Calculate V_{out} from $V_{out} = A_{v(tot)} V_{in}$.

    d. Calculate I_{in} from $I_{in} = V_{in} / R_{in(tot)}$. Use the calculated value of $R_{in(tot)}$ from Table 5-2.

    e. Calculate I_{out} from $I_{out} = V_{out} / R_L$.

    f. Calculate the current gain from $A_{i(tot)} = I_{out} / I_{in}$.

    g. Calculate the power gain from $A_p = A_{v(tot)} A_{i(tot)}$.

2) Use the oscilloscope to measure V_{in} and V_{out} and calculate $A_{v(tot)}$ from the measured voltages. Use the previously measured value of $R_{in(tot)}$ to calculate the input current and determine the measured output current by applying Ohm's law to the measured V_{out}. Then use the measured voltage and current gains to measure the power gain. Complete Table 5-5, showing three significant digits for all entries.

Table 5-5: Cascaded Amplifier AC Values

AC Quantity	Computed Value	Measured Value
V_{in}	150 mV$_p$	
$A_{v(tot)}$		
V_{out}		
I_{in}		
I_{out}		
$A_{i(tot)}$		
A_p		

Conclusion for Part 2

Questions for Part 2

1) The power gain for a common-collector (CC) amplifier is approximately the same as the current gain. Explain why.

2) How would you add a gain control to the two-transistor amplifier such that there was no effect on dc quantities?

3) What factors would contribute to the discrepancy between the calculated and measured power gains for the cascaded amplifier?

Devices Experiment 6 - Power Amplifiers

6.1 Introduction

Power amplifiers are amplifiers that must be able to provide power to drive a load – anywhere from a few watts to hundreds of watts. Heat dissipation and efficiency are always considerations with power amplifiers.

Except in low-power applications, the class-A amplifier is not widely used because it is not particularly efficient. In Part 1 of this experiment, you will measure the power gain and efficiency of a basic class-A power amplifier. The amplifier you will test uses a Darlington arrangement of transistors to improve the power gain. The total power is generally low to avoid excessive heat.

In Part 2, you will test a class-AB (push-pull) amplifier that includes a voltage gain stage as part of the circuit. The class-AB amplifier is much more efficient than the class-A amplifier, so it is widely used when more power is needed. Class-AB amplifiers use a *diode current-mirror* to bias the transistors into slight conduction and avoid a specific distortion called *cross-over distortion*. The forward-biased diodes will each have approximately the same 0.7 V drop as the base-emitter junction, so the quiescent current in the transistors will be nearly the same as the diode current (depending on how well the bias diodes match the base-emitter diodes).

6.2 Reading

Floyd, *Electronic Devices, 8th Edition*, Chapter 7

6.3 Key Objectives

Part 1: Calculate and measure the dc and ac characteristics for a class-A CC amplifier including the power gain and the efficiency.

Part 2: Calculate and measure the dc and ac characteristics for an amplifier that includes a class-AB stage.

6.4 Multisim Files

Part 1: *Devices_Exp_06_Part_01*

Part 2: *Devices_Exp_06_Part_02*

Part 1: The Class A Power Amplifier

The circuit in this part is a class-A power amplifier that is run from a 12 V supply.

1) Open the Multisim file *Devices_Exp_06_Part_01*.

2) Calculate the dc parameters listed in Table 6-1. Enter each value in the "Calculated Value" column.

    a. Assume the input voltage divider is stiff. This is a reasonable assumption because the equivalent beta of the Darlington arrangement is very high. Use an unloaded voltage divider to calculate the base voltage, V_B.

    b. There are two forward drops between the base and the emitter of Q_2. Calculate the emitter voltage, V_E, taking this into account.

    c. Apply Ohm's law to R_E to calculate I_E.

    d. Calculate V_{CE} for Q_2.

3) Reduce V_S to 0 V. This ensures that the ac source will not affect the dc measurements. Measure each of the quantities listed in Table 6-1 and record them in the "Measured Value" column. An ammeter is already in place to measure the collector current.

Table 6-1: Power Amplifier DC Values

CC Amp ($Q_{1,2}$)	Calculated Value	Measured Value
V_B		
V_E		
$I_E \approx I_C$		
V_{CE}		

4) In this step, you will calculate the input resistance and the power gain of the amplifier. Enter the calculated values in Table 6-2.

    a. Calculate the input resistance, R_{in} for the amplifier. The input resistance to the base of Q_2 is much larger than the bias resistors due to the Darlington arrangement. To simplify the calculation, assume only the bias resistors determine R_{in}.

    b. The value of the load resistor, R_L given in the Multisim experiment is shown in Table 6-2 as the calculated value for R_L.

    c. The voltage gain of a CC amplifier is always less than 1, but to calculate it in detail requires that each transistor be treated separately and it is necessary to know β accurately. For a practical circuit, it is reasonable to estimate the overall voltage gain as 0.95, realizing that the measured value may be different. This assumed value is shown in Table 6-2.

    d. The power gain is given by

$$A_p = A_v^2(R_{in} / R_L)$$

Calculate the power gain using the values in Table 6-2. Enter the calculated value in Table 6-2.

Table 6-2: Computed Values for Class AB Power Amplifier

Quantity	Calculated Value
Input resistance, R_{in}	
Load resistance, R_L	
Voltage gain, A_v	
Power gain, A_p	

5) In this step, you will start by measuring the input resistance of the amplifier, the voltage gain, and then use the same equation given in step 4d to determine the measured power gain Set up the source voltage for a 2.5 V_p input. Observe the input and output signals using the oscilloscope.

    a. Set the value of R_{test} to the computed value of R_{in} and measure the output voltage. Adjust R_{test} as needed until the output is half the input. R_{test} will then be equal to the input resistance. Record this as the measured R_{in} in Table 6-3. After recording the value, change the test resistor back to 0 Ω.

    b. As before, the value of the load resistor is shown in Table 6-3. In a practical lab experiment, this would be measured with an ohmmeter, but for this simulation the nominal value will be used.

    c. Measure the output voltage across the load (V_{out}) and determine the measured voltage gain by calculating the ratio of V_{out}/V_{in}. Record the measured voltage gain, A_v, in Table 6-3.

    d. Using the measured value of V_{out}, determine the measured output power from $A_p = A_v^2(R_{in} / R_L)$. Record this value in Table 6-3.

Table 6-3: Measured Values for Class AB Power Amplifier

Quantity	Measured Value
Input resistance, R_{in}	
Load resistance, R_L	
Voltage gain, A_v	
Power gain, A_p	

6) Calculate the efficiency of the circuit. First calculate P_{out} based on the ac power delivered to the load. Then measure the total dc current from the power supply by turning off the circuit and inserting an ammeter in the supply line. Use this value and V_{CC} to calculate the total dc power. Calculate the efficiency by $\eta = P_{out} / P_{DC}$. Show your work:

7) Try experimenting with the circuit by changing different values and observing the results. What happens if the power supply voltage is increased or the load resistor decreased? What happens if the signal is increased?

   Observations:

Conclusions for Part 1

Questions for Part 1

1) R_E is specified as a 2 W resistor. If there is no input signal, could a smaller wattage resistor be used? Explain your answer.

2) With no input signal, how much power was dissipated in Q_2? Do you think it is necessary to use a heat sink with this transistor?

3) Power gain is the ratio of power delivered to the load (P_{out}) to the power supplied to the amplifier (P_{in}). Using your measured quantities, calculate the rms values of P_{out} and P_{in} for the experiment and show that the power gain is the same as your calculated value in step 5. Show your work.

Part 2:   The Class AB Power Amplifier

1) Open the Multisim file *Devices_Exp_06_Part_02*. The circuit includes a class-AB power amplifier using complementary BD329 and BD330 transistors. These transistors are designed for use in portable power applications. You will first calculate dc and ac parameters for the circuit and then measure the parameters in Multisim. Start with the dc parameters as follows:

   a. The method described in this step starts by calculating a Thevenin circuit for the input in order to obtain a reasonably good value for the emitter current, $I_{E(Q3)}$. The following steps will guide you:

      i.   Mentally break the circuit at the base of Q_3 and look back toward the source. The Thevenin circuit will have $R_1 \| R_2$, which is the Thevenin resistance, R_{TH}.

      ii. The Thevenin voltage, which is negative in this case, is the unloaded base voltage:

      $$V_{TH} = V_{EE} [R_1 / (R_1 + R_2)]$$

      iii. Next write Kirchhoff's voltage law around the loop consisting of the Thevenin circuit and the base-emitter circuit for Q_3, which include R_4, R_5, and V_{EE} (be careful of polarities).

      iv. From the result, calculate $I_{E(Q3)}$ and enter the calculated value in Table 6-4. The current is somewhat dependent on β_{DC}. Assume a value of 150 for this calculation. Writing the equation is left as an exercise, but the first step is completed to help you get started. Show your work in the space provided:

      $$V_{TH} + I_B R_{TH} + V_{BE} + I_{E(Q3)} R_4 + I_{E(Q3)} R_5 - V_{EE} = 0$$

   b. Calculate $V_{(R4+R5)}$ by calculating the voltage drop across both R_4 and R_5 using Ohm's law.

   c. Calculate $V_{E(Q3)}$ by algebraically adding $V_{(R4+R5)}$ to V_{EE}.

   d. Calculate $V_{B(Q1)}$ by applying Ohm's law to R_3 and subtracting the result from V_{CC}.

   e. Calculate $V_{B(Q2)}$ by subtracting two diode drops from $V_{B(Q1)}$. Note that this is also $V_{C(Q3)}$.

   f. Calculate $V_{CE(Q3)}$ by subtracting $V_{E(Q3)}$ from $V_{C(Q3)}$.

2) Turn on power to the circuit. V_S should be at 0 V so as not to affect the dc readings. Measure the dc quantities listed in Table 6-4 with a dc voltmeter. Measure I_E indirectly by measuring V_{R3} and applying Ohm's law. Record your readings using 3 significant digits in the "Measured Value" column of Table 6-4. Your measured values may vary somewhat from the calculated values but should be reasonably close. If they are not, recheck your calculations.

Table 6-4: Class AB Power Amplifier DC Values

DC Quantity	Computed Value	Measured Value
$I_{E(Q3)} \approx I_{C(Q3)}$		
$V_{(R3+R4)}$		
$V_{E(Q3)}$		
$V_{B(Q1)}$		
$V_{B(Q2)} = V_{C(Q3)}$		
$V_{CE(Q3)}$		

3) Calculate and record the ac quantities listed in Table 6-5. Steps for the calculations are as follows:

a. Set the input voltage to 1.414 V_p (1.0 V_{rms}). (This value is entered as the computed value.)

b. Calculate $r'_{e(Q3)}$ from $r'_{e(Q3)} = 25$ mV / $I_{E(Q3)}$.

c. Assuming the potentiometer is set midway ($0.5R_4$), calculate the voltage gain $A_{v(Q3)}$ for Q_3 from

$$A_{v(Q3)} = (R_3 \| \beta_{ac(Q1)}) / (r'_{e(Q3)} + R_3 + 0.5R_4)$$

Assume a typical $\beta_{ac(Q1)} = 150$.

d. Assume a voltage gain $A_{v(Q1,Q2)}$ for the push-pull stages of 0.95.

e. Calculate $A_{v(overall)}$ by multiplying $A_{v(Q3)}$ by $A_{v(Q1,Q2)}$.

f. Multiply V_{in} by the computed voltage gain to calculate V_{out}.

g. Calculate the power P_{out} delivered to the load for this particular input using V_{out}^2 / R_L.

4) Using the oscilloscope, measure the quantities given in Table 6-5 (except $r'_{e(Q3)}$). Use the measured output voltage to calculate the output power, and enter this as the measured value.

5) While viewing the output signal, vary R_5 by typing "A" or "Shift + A".

Observations:

Table 6-5: Class AB Power Amplifier AC Values

AC Quantity	Computed Value	Measured Value
$V_{in} = V_b$	1.414 V_p	
$R'_{e(Q3)}$		
$A_{v(Q3)}$		
$A_{v(Q1,Q2)}$		
$A_{v(overall)}$		
V_{out}		
P_{out}		

6) Determine the power gain for the amplifier. Summarize the method you used to determine the power gain.

7) Insert a short from the base of Q_1 to the base of Q_2.

Observations

Conclusions for Part 2

Questions for Part 2

1) Explain how you would measure the overall maximum efficiency of the amplifier.

2) What symptoms would you expect to see if Q_2 was open?

Devices Experiment 7 - JFET Characteristics

7.1 Introduction

Field-effect transistors (FETs) were actually thought of long before the invention of the BJT. Dr Julius Lilienfeld of Germany applied for a patent in Canada in 1925 and later in the U.S. for a FET device. However, it wasn't until the 1960s that FETs became commercially available, a decade after BJTs were available.

FETs work on an entirely different principle than BJTs. They are voltage-controlled devices rather than current-controlled devices as in the case of BJTs, and they can be thought of as a normally-on device (as opposed to a BJT, which needs current to turn on). For the characteristic curve, the maximum drain current occurs when the input (gate voltage) is zero.

In Part 1 of the experiment, you will plot the characteristic curve for a small JFET, which is a plot of drain source voltage as a function of drain current for a given gate source voltage. The JFET selected for this part is the BF410B, which has characteristics that are useful for an application that is in Part 2. This particular JFET is one of a family of four JFETS with varying I_{DSS} specifications and is useful in various communications devices because of its low noise and ability to operate at high frequencies. (The other JFETS in this family are the BF410A, BF410C, and BF410D.)

Part 2 introduces a circuit that uses the ohmic region of the JFET as a variable resistor to control the gain of the basic amplifier (with some modifications) that was introduced in Part 2 of Experiment 5. The circuit is designed to illustrate automatic gain control (AGC), so that if the input *increases* beyond the normal range, the gain *decreases* to compensate. The ohmic region is near $V_{DS} = 0$, and for this part, the operating point is set at $V_{DS} = 0$ by connecting a capacitor in series with the drain. AGC circuits are common in high-frequency communication applications. Usually they are applied to several stages of an amplifier, but for purpose of seeing how they work, we will use a simpler version applied to only one stage.

7.2 Reading

Floyd, *Electronic Devices, 8th Edition*, Chapter 8

7.3 Key Objectives

Part 1:   Measure and plot the characteristic curves for an *n*-channel JFET.

Part 2:   Test a modified amplifier from Experiment 5 using a JFET as a voltage-controlled resistor to automatically control gain.

7.4 Multisim Files

Part 1:   *Devices_Exp_07_Part_01*

Part 2:   *Devices_Exp_07_Part_02*

Part 1:   JFET Characteristic Curve

The circuit in this part is similar to the one in Part 2 of Experiment 5. The principal difference is that this circuit uses negative voltage sources for the control signals because you are testing an *n*-channel JFET.

1) Open the Multisim file *Devices_Exp_07_Part_01*. Connect the Tektronix oscilloscope so that channel 1 is connected to the drain of the transistor and channel 2 is connected to the top of R_D.

2) Turn on the oscilloscope.

3) In the **DISPLAY** menu, choose **Format XY**.

4) In the **CH 1 MENU** choose **Coupling Ground**. This disconnects the *x*-input so that you can see where ground is positioned on the *x*-axis.

5) Activate the circuit. You should see a vertical line near the center of the screen. Use the **CH 1 POSITION** control to move the line to the left side of the screen in line with the first vertical line on the display. This will cause all divisions on the *x*-axis to have positive values starting from the left side.

6) In the **CH 1 MENU** choose **Coupling DC**. (This reconnects the channel to the input.)

7) Set the Channel 1 **VOLTS/DIV** control to "500 mV/Div".

8) In the **CH 2 MENU** choose **Coupling Ground**. This disconnects the *y*-input enabling you to see where ground is positioned on the *y*-axis.

9) Move the **CH 2 POSITION** control until the line is positioned along the bottom of the screen. This will cause all divisions on the *y*-axis to have positive values.

10) In the **CH 2 MENU** choose **Coupling DC** and **Invert On**. This causes the increase in current to be displayed in the positive direction.

11) Set the Channel 2 **VOLTS/DIV** control to "500 mV/DIV".

12) You should observe one of the characteristic curves for the transistor. You can change the gate voltage by selecting different switch positions ("A", "B", "C") on the input. When ground is selected for the gate voltage, the current will be I_{DSS}.

13) Sketch the four characteristic curves on Figure 7-1 for the four different gate voltages. Label the graph and identify each curve. In particular, since Part 2 uses the ohmic region, notice how the curves change near the origin.

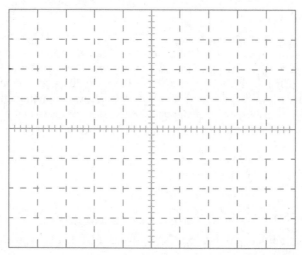

Figure 7-1: Characteristic Curves for a BF410B JFET

Observations:

14) Substitute a BF410D for the BF410B by double-clicking the JFET and using the **Replace** button. Decide what changes you need to make to the circuit and the scope settings to show the characteristic curves for the BF410D. List changes you made and sketch the characteristics for the BF410D in Figure 7-2. Label your sketch.

Changes to the circuit and scope:

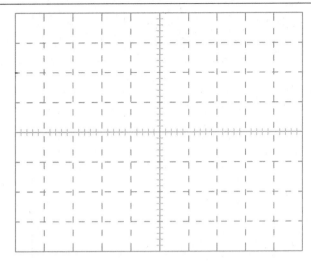

Figure 7-2: Characteristic Curves for a BF410D JFET

Conclusions for Part 1

Questions for Part 1

1) Based on your data, what are the approximate values of I_{DSS} for each transistor?

2) Based on your data, what is $V_{GS(off)}$ for each transistor?

Part 2: Automatic Gain Control with a JFET

1) Open the Multisim file *Devices_Exp_07_Part_02*. The circuit is essentially the same as the 2-stage amplifier from Part 2 of Experiment 5 (with some modifications). The biggest change is to add the 2N410B in the emitter circuit of Q_1 (and change the emitter resistance). The circuit also connects a voltmeter to the gate of the JFET so that you can monitor the gate voltage as the input is varied. Connect the oscilloscope so that channel 1 monitors the input and channel 2 monitors the output. Activate the circuit and adjust the scope to view the input and output.

2) The initial input voltage is set to 100 mV peak. Using the oscilloscope, measure the peak output voltage $V_{out(peak)}$ and record it in Table 7-1.

3) Set $V_{in(peak)}$ to each value in Table 7-1. Measure and record each value of $V_{out(peak)}$ in Table 7-1. You may need to wait several seconds for Multisim to settle to a final value.

4) Use $V_{out(peak)}$ and $V_{in(peak)}$ to calculate the voltage gain, A_v for each setting. Record your data in Table 7-1. Notice what happens to the gate voltage as you change settings.

Table 7-1: JFET Automatic Gain Control Data

$V_{in(peak)}$	$V_{out(peak)}$	Voltage Gain, A_v
100 mV		
200 mV		
300 mV		
400 mV		
500 mV		
600 mV		
700 mV		
800 mV		
900 mV		
1.0 V		
1.1 V		

Observations:

5) Plot the voltage gain as a function of the input voltage in Plot 7-1.

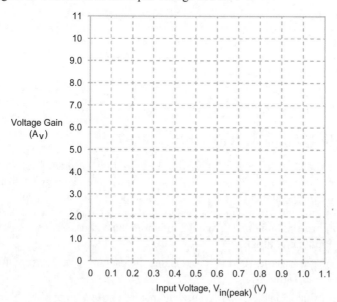

Plot 7-1: JFET Amplifier Gain versus Input Voltage

Conclusions for Part 2

Questions for Part 2

1) How does the automatic gain control generate a negative voltage to the gate of the FET?

2) Where is the Q-point for the JFET in this circuit?

Devices Experiment 8 - FET Amplifiers

8.1   Introduction

Field-effect transistors give designers important advantages for linear amplifiers. The most important advantages are very high input resistance and low noise. This is an advantage for situations like amplifying low level signals such as found in communication circuits.

This experiment focuses on two important types of FET amplifier – the common-source (CS) amplifier and the common-drain (CD) amplifier. These designations refer to ac parameters. For the CS amplifier, the source is the common terminal to ac, and for the CD amplifier the drain is the common terminal.

In Part 1 of the experiment, you will investigate the CS amplifier with a combination of voltage-divider bias and self-bias. As in most electronic circuits, there are advantages (more stability) and disadvantages (more components) to this method. The first step illustrates the design steps in detail. Follow the procedure to see how the selected values were chosen. After this, you will investigate a CD amplifier using the same JFET. You will start with a traditional CD amplifier, which has certain limitations as you will see. A much better version of the CD amplifier is then tested using current-source biasing.

Part 2 uses the CS amplifier from Part 1 as a first stage in a complete amplifier that includes BJT circuits with which you should already be familiar. The dc and ac parameters are left for you to calculate and compare with the Multisim measured results.

8.2   Reading

Floyd, *Electronic Devices, 8th Edition*, Chapter 9

8.3   Key Objectives

Part 1:   Calculate and measure dc and ac parameters for a common-source (CS) amplifier and for two common drain (CD) amplifiers.

Part 2:   Test a complete amplifier consisting of a FET input stage, a BJT voltage gain stage, and a CC output stage.

8.4   Multisim Files

Part 1:   *Devices_Exp_08_Part_01a, Devices_Exp_08_Part_01b*, and *Devices_Exp_08_Part_01c*

Part 2:   *Devices_Exp_08_Part_02*

Part 1:   JFET Amplifiers

8.5   Common-Source Amplifier

1)   Open the Multisim file *Devices_Exp_08_Part_01a*. In this first step, a method for choosing a reasonable bias resistor is given, with the following steps. The circuit uses a combination of voltage-divider bias (with the gate at a positive voltage) and self-bias. This source voltage will be more positive than the gate due to the self-bias resistor, so it meets the requirement that the gate is negative with respect to the source. The advantage of this method is there is less change in the circuit parameters (including the gain) if the transistor is replaced with another one with different specifications.

      a.   Choose a JFET based on the requirements for the circuit. For purpose of illustration, the BF545B is a general-purpose JFET that can be used with basic amplifiers. To illustrate voltage-divider bias in combination with self-bias, choose a small positive gate voltage to start. For this step, a 1.0 MΩ resistor to ground and a 10 MΩ resistor to +15 V will produce the small positive gate voltage needed for this bias method. (Note that if you use only self-bias, the gate is at 0 V.)

      b.   To choose a reasonable self-bias resistor value, it is helpful to view the range of transconductance curve, based on the manufacturer's specified minimum and maximum transconductance curves. A resistor line is started in the first quadrant at the gate voltage (in this case 1.2 V) and extended into the second quadrant, intersecting the minimum and maximum transconductance curves for the BF545B, as shown in Figure 8-1. This curve shows that a 680 Ω resistor is a reasonable choice for a self bias resistor because it intersects the minimum transconductance curve near the center.

c. Finally, choose a drain resistor that drops about 1/3 of V_{DD} across the transistor. This will cause the Q-point on the ac load line to be near the center. From the transconductance curves, it can be seen that the drain current will be about 3 mA (but is dependent on the transistor.) A 2.7 kΩ resistor drain is a good choice because it will provide about 1/3 of V_{DD} across the transistor. The circuit design is completed with the addition of a load resistor that will simulate the input resistance of a following stage.

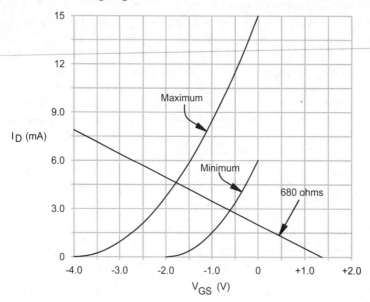

Figure 8-1: Transconductance Curves for the BF545B JFET

2) Calculate the dc parameters for the amplifier circuit. The input current is very small for a JFET, so it is reasonable to ignore any loading effect. Based on the transconductance curve and the crossing points for a 680 Ω resistor, the current indicated by the graph in Figure 8-1 will be approximately 3.0 mA; because of the variation between FETs, this is necessarily an approximation. Complete Table 8-1, showing the calculated values based on this assumed current. The value for $I_S = I_D$ is already shown.

3) With the input ac source voltage set to zero, use a voltmeter to measure and record the dc parameters in Table 8-1. To measure the current, measure the voltage across R_D and apply Ohm's law.

Table 8-1: JFET Common Source Amplifier DC Parameters

DC Parameter	Computed Value	Measured Value
V_G		
$I_S \approx I_D$	3.0 mA	
V_S		
V_{RD}		
V_D		
V_{DS}		

4) Because of variations in FETs, the ac parameters will also be approximate. To estimate the voltage gain, you need to know the transconductance, g_m. You can estimate it from the curves, looking for a change in drain current and dividing by an equivalent change in the gate-source voltage. A triangle is drawn on the curve as illustrated in Figure 8-2 (showing a portion of the transconductance curve.) Notice that a change of 3.05 mA in I_D corresponds to a 0.90 V change in V_{GS}. From this we can deduce that

$$g_m \approx \Delta I_D / \Delta V_{GS} = 3.05 \text{ mA} / 0.90 \text{ V} = 3.39 \text{ mS}$$

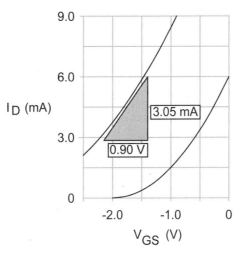

Figure 8-2: Reading the Transconductance Curve near the Q-point.

5) Calculate the ac parameters given in Table 8-2.

    a. Calculate the voltage gain from $A_v = g_m R_d$. Use the parallel combination of the drain resistor and load resistor for the ac drain resistance, R_d. (This is the loaded gain).

    b. Assuming the input is set for 50 mV$_p$, calculate the output voltage, V_{out}.

    c. Calculate R_{in}, assuming that the JFET has infinite input resistance.

Table 8-2: JFET Common-Source Amplifier AC Parameters

AC Parameter	Computed Value	Measured Value
A_v		
V_{out}		
R_{in}		

6) Measure and record the parameters in Table 8-2. To measure the input resistance, start by setting R_{test} to the calculated value and then adjust it until the output drops by exactly one-half of its original value.

8.6 Common-Drain Amplifier

The common-drain (CD) configuration is used in cases where the signal amplitude does not need to be increased because, like the CC amplifier, the voltage gain is always less than 1.

1) Open the Multisim file *Devices_Exp_08_Part_01b*. This circuit is a traditional common-drain amplifier.

2) Connect the oscilloscope and observe the input and output signals. Measure the voltage gain with no load.

3) Connect the 10 kΩ load and observe what happens to the gain. Is there any evidence of distortion?

4) Raise the input signal and observe what happens.

5) Summarize your observations.

8.7 Common-Drain Amplifier with Current-Source Biasing

A much better common-drain circuit can be constructed by using current-source biasing. Recall that an ideal current source has infinite input resistance. A JFET can approach this idea for this application.

1) Open the Multisim file *Devices_Exp_08_Part_01c*. This circuit uses a second JFET (Q_2) to act as a current source load for the first one. An advantage of this circuit is that the output is directly coupled (no capacitor needed).

2) Connect the oscilloscope and observe the input and output signals. Note that the output is taken from the drain of Q_2. Measure the voltage gain with no load.

3) Connect the 10 kΩ load and observe what happens to the gain. Is there any evidence of distortion?

4) Raise the input signal and observe what happens.

5) Adjust R_{S2} (using the "A" key) and note what happens.

6) Summarize your observations.

Conclusions for Part 1

Questions for Part 1

1) With the combination voltage-divider and self-bias in the CS amplifier, a larger source resistor can be used to provide the same source current as compared to self-bias by itself. What advantage does this offer for practical circuits?

2) In the CS amplifier circuit, increasing R_D can increase the gain. What is the disadvantage of doing this?

3) In the CD amplifier with current-source biasing, what do you think is the purpose of R_{S2}?

Part 2: A Complete Amplifier

1) Open the Multisim file *Devices_Exp_08_Part_02*. This circuit has the CS amplifier from Part 1 for the input stage. This design uses a mix of FETs and BJTs because, in general, FETs are better suited for input stages, where low noise and high input impedance are important, and BJTs are more suited where voltage gain is important.

2) From your knowledge of solving for dc conditions, calculate the dc conditions listed in Table 8-3 and enter your calculated values in the table. Assume that voltage-dividers are stiff (unloaded). For the measured current, measure the voltage across a resistor and apply Ohm's law, which is the procedure you will normally use in practical work.

3) Measure the dc parameters. Set the signal source to zero to avoid any effect from converting ac to dc and use the DMM to measure voltages. Enter the measured values in Table 8-3. You should see reasonable agreement with your calculated values.

Table 8-3: Complete Amplifier DC Parameters

DC Parameter	Computed Value	Measured Value
$V_{B(Q2)}$		
$V_{EQ2)}$		
$I_{E(Q2)} \approx I_{C(Q2)}$		
V_{RC}		
$V_{C(Q2)} = V_{B(Q3)}$		
$V_{E(Q4)}$		

4) Calculate the ac parameters listed in Table 8-4.

   a. Assume the input voltage is set to 50 mV$_p$ and the calculated gain of the first stage ($A_{v(Q1)}$) is the same as the loaded gain calculated in Table 8-2.

   b. Calculate the ac drain voltage from the input and gain.

   c. For $r'_{e(Q2)}$, use $r'_{e(Q2)} = 25$ mV $/ I_{E(Q2)}$.

   d. Calculate the gain of Q_2 by $A_{v(Q2)} = R_C/R_{e(Q2)}$. The Darlington transistors have very high β, so loading effects on Q_2 can be ignored.

   e. Calculate the collector voltage of Q_2, $V_{c(Q2)}$, by multiplying $A_{v(Q2)}$ by the input voltage to Q_2, which is the ac drain voltage of Q_1, $V_{d(Q1)}$

   f. Assume the gain of the Darlington CC amplifier ($A_{v(Q2.3)}$) is 0.95. Calculate the overall gain from $A_{v(overall)} = (A_{v(Q1)}) (A_{v(Q2)}) (A_{v(Q3,4)})$

   g. Calculate the output voltage by multiplying the input voltage by the overall gain.

5) Set the input voltage to 50 mV$_p$. Using the oscilloscope, measure and record the parameters listed in Table 8-4.

Table 8-4: Combined Amplifier AC Parameters

AC Parameter	Computed Value	Measured Value
$A_{v(Q1)}$		
$V_{d(Q1)}$		
$r'_{e(Q2)}$		
$A_{v(Q2)}$		
$V_{c(Q2)}$		
$A_{v(overall)}$		
V_{out}		

Observations

Conclusions for Part 2

Questions for Part 2

1) How would you modify the circuit to add a gain control? Explain the circuit change in detail.

2) From your measured data, what dc power is dissipated in Q_4 and R_{E3}?

Devices Experiment 9 - Amplifier Frequency Response

9.1 Introduction

The frequency response of transistor amplifiers is affected by the reactive components in the circuit. In a typical BJT amplifier, the low-frequency response is determined by the coupling and bypass capacitors. The analysis of the low-frequency response is done by considering each capacitor's discharge path and forming an equivalent (and simplified) *RC* circuit.

In Part 1 of this experiment, you will analyze a high-gain CE amplifier. The low-frequency response is calculated based on the charge/discharge paths for the coupling and bypass capacitors. Keep in mind that the path is drawn from the perspective of the capacitor. The process is illustrated in this experiment; you should follow the steps in finding the equivalent resistance. The low-frequency response of each capacitor is then measured by isolating the capacitor.

In Part 2, you will investigate the upper critical frequency for the same amplifier. Normally, the internal junction capacitances, chiefly the base-emitter capacitance and base-collector capacitance, control the high frequency response. These internal capacitances are proportional to the physical area of the junctions and inversely proportional to the width of the depletion region. The base-emitter capacitance is not constant, but rises with increasing bias current due to charge depletion in the junction. The base-collector junction is reverse biased so the capacitance is small due to a wider depletion region, but the Miller effect can act as a multiplier to this small capacitance for inverting amplifiers. With a high gain inverting amplifier, the Miller effect is dominant and controls the high-frequency response. In order to provide a simple method for you to control the effect of the base-collector capacitance and to reduce the upper cutoff frequency, an additional capacitor is connected in parallel with the base-collector junction in this experiment. This will cause the upper cutoff frequency to be lower, and it will simplify the measurements if you choose to build it in the lab. Again, the key to analysis is to form a simplified equivalent circuit and use it to calculate the response.

9.2 Reading

Floyd, *Electronic Devices, 8th Edition*, Chapter 10

9.3 Key Objectives

Part 1: Compute and measure the three lower critical frequencies for a CE amplifier for the coupling and bypass capacitors. Measure the overall f_{cl} with the oscilloscope and the Bode plotter.

Part 2: Compute the two upper critical frequencies for a CE amplifier and use them to compute the overall upper critical frequency. Measure the overall f_{cu} with the oscilloscope and the Bode plotter.

9.4 Multisim Files

Part 1: *Devices_Exp_09_Part_01*

Part 2: *Devices_Exp_09_Part_02*

Part 1:   Low-Frequency Response

1) Open the Multisim file *Devices_Exp_09_Part_01*.

2) Calculate and record the dc parameters given in Table 9-1 for the circuit. To get an accurate calculation, thevenize the input bias circuit. After writing Kirchhoff's voltage law, you can find that

$$I_E = \frac{V_{TH} - V_{BE}}{R_{E1} + R_{E2} + \dfrac{R_{TH}}{\beta_{DC}}}$$

Apply this equation to calculate I_E, which is approximately I_C.

3) Use Ohm's law to calculate V_E and V_C. Record these values in Table 9-1.

4) Calculate V_{CE} by subtracting V_E from V_C. Record these values in Table 9-1.

5) With the input ac source voltage set to zero, use a DMM to measure the parameters in Table 9-1. To measure I_C, measure the voltage across R_C and apply Ohm's law.

Table 9-1: Amplifier DC Parameters

DC Parameter	Computed Value	Measured Value
$I_C \approx I_E$		
V_E		
V_C		
V_{CE}		

6) Calculate the ac parameters listed in Table 9-2. The input voltage is shown as 20 mV$_p$. Use this value to calculate and record the output voltage, V_{out}.

7) Measure and record the parameters (except r'_e) in Table 9-2.

Table 9-2: Amplifier AC Parameters

AC Parameter	Computed Value	Measured Value
$V_{in} = V_b$	20 mV$_p$	
r'_e		
A_v		
$V_{out} = V_c$		

8) The lower critical frequency, f_{cl}, is due to the coupling capacitors, the emitter bypass capacitor and their associated resistive networks that provide a charge and discharge path. To calculate f_{cl} for each capacitor, you need to analyze this path and form an equivalent single RC circuit for each one. Steps are:

    a. C_1: Tracing the path for C_1, you see several paths to ground, as illustrated in Figure 9-1 with the dotted lines. (Recall that the power supply is an ac ground.)

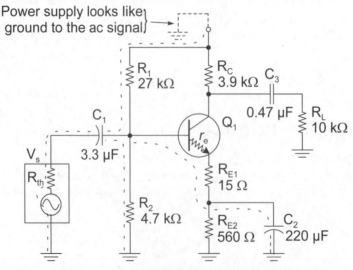

Figure 9-1: Circuit Paths to Ground for C_1

On the right side of C_1 are the bias resistors (R_1 and R_2) and the ac resistance of the emitter circuit consisting of $R_{E1} + r'_e$. Together, these resistors are equivalent to a single resistance, R_{in}. On the left side of C_1 you see only the Thevenin resistance of the source, R_{th}). Thus, the total equivalent resistance as seen by C_1 is:

$$R_{eq} = R_{in} + R_{th}$$
$$= (\beta_{ac}(R_{E1} + r'_e) \| R_1 \| R_2) + R_{th}$$

b. Using this equation, compute the equivalent resistance seen by C_1. For β_{ac}, assume 170. The Thevenin source resistance of the source is 50 Ω. Enter the computed value of R_{eq} in Table 9-3.

c. C_2: For C_2, R_{E2} is one of two parallel paths. The other parallel path is the combination of $R_{E1} + r'_e$ plus the reflected resistance of the base circuit, which can be ignored. (The reflected resistance of the base circuit less than 1 Ω because the resistance looking back into the base circuit is divided by β_{ac} when you are looking for the equivalent resistance in the emitter circuit.) Thus the equivalent resistance for C_2 is:

$$R_{eq} = R_{E2} \| (R_{E1} + r'_e)$$

Using this equation, compute the equivalent resistance R_{eq} seen by C_2 and enter the computed value in Table 9-3.

d. C_3: This path is simply R_C in *series* with R_L.

Table 9-3: Equivalent Resistance R_{eq} for Capacitors

Capacitor	Capacitor Value	R_{eq}
C_1	3.3 μF	
C_2	220 μF	
C_3	0.47 μF	

9) Compute the lower critical frequency f_{cl} for each capacitor (C_1, C_2, and C_3) from the equation:

$$f_{cl} = 1 / (2\pi R_{eq}C)$$

Use the R_{eq} from Table 9-3 for each capacitor. Enter the computed lower critical frequency for each capacitor in Table 9-4. The overall critical frequency of the amplifier will be higher than the highest frequency determined for each individual capacitor. The actual total is dependent on the interactions but will be lower than the sum of the various cutoff frequencies.

10) Measure the lower cutoff frequency f_{cl} due to each capacitor. To accomplish this, you need to make the effect of only one capacitor dominant and measure its cutoff frequency. In an actual experiment, you can do this by "swamping" out the effect of two of the three capacitors by placing a very large capacitor in parallel with them and measuring the overall response; this causes the overall frequency response to be determined by the one capacitor that is not "swamped". In Multisim, you can simply increase the capacitance of two of the capacitors by a large amount.

a. C_1: Increase the capacitance of C_2 and C_3 by a factor of 100. Then read the response by using the oscilloscope and checking with the Bode plotter. Your readings should be close to the same in both cases. Steps are as follows:

    i. Oscilloscope measurement: Observe the output signal in midband (around 10 kHz) and measure the amplitude of the signal. Decrease the frequency until the output drops to 70.7% of the voltage in midband. This frequency is the lower critical frequency due to C_1. Measure and record the value in Table 9-4.

    ii. Bode plotter measurement: Connect the **IN** line to the top of V_S (because the Bode plotter needs a fixed reference.) Connect the **OUT** line to the output and connect the – inputs to ground. Set the mode to **MAGNITUDE** and both horizontal and vertical axes controls to **LOG**. On the horizontal axis, set the initial value (**I**) to "10 Hz" and the final value (**F**) to "100 kHz". On the vertical axis, set the initial value (**I**) to "10 dB" and the final value (**F**) to "50 dB". These settings will enable you to see the lower frequency response with reasonable resolution. Drag the cursor to the right so that you are well into the flat portion of the response. Note the decibel reading, which is the gain at midband. Move the cursor to the left, watching the decibel response until it is 3 dB less than the gain at midband. Read the frequency; this is the cutoff frequency. Record the value in Table 9-4.

b. C_2: Restore C_2 to its original value and increase the capacitance of C_1 by a factor of 100. Repeat the procedure for measuring the response with the oscilloscope and the Bode plotter described previously. Record the value in Table 9-4.

c. C_3: Restore C_3 to its original value and increase the capacitance of C_2 by a factor of 100. Repeat the procedure for measuring the response with the oscilloscope and the Bode plotter described previously. Record the value in Table 9-4.

Table 9-4: Measured Lower Cutoff Frequencies

Capacitor	Calculated f_{cl}	Measured f_{cl}
C_1		
C_2		
C_3		

11) Restore all capacitors to their original values. Measure the overall response and summarize your observations.

Conclusions for Part 1

Questions for Part 1

1) Why does the bypass capacitor (C_2) have the largest value?

2) If the circuit designer wanted to make the bias stiffer by choosing smaller bias resistors, would there be an effect on the frequency response? Explain your answer.

3) If a JFET had been used instead of a BJT, would there be any reason to change the input coupling capacitor's value? Explain your answer.

Part 2:   High-Frequency Response

1) Open the Multisim file *Devices_Exp_09_Part_02*. This circuit is the same as in Part 1, with an additional capacitor. The purpose of the additional capacitor (C_4) is to provide a simple method for you to easily

change the capacitance between the base and collector and observe the effect on the upper frequency response. In practical circuits a physical capacitor is generally not used here unless the designer specifically wants to limit the high-frequency response, but the small base-collector capacitance is actually more important in inverting amplifiers than the larger base-emitter capacitance.

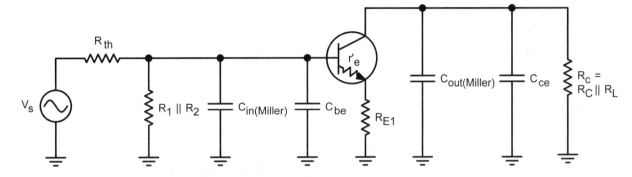

Figure 9-2: High-Frequency Equivalent Circuit

2) If you did not do Part 1, you will need to compute the ac and dc parameters for the CE amplifier and complete Tables 9-1 and 9-2. You should complete steps 1 to 7 in Part 1 before continuing.

3) In this step, you will compute the critical frequency due to the input network. Record the computed values in Table 9-5.

    a. Determine the equivalent input capacitance, C_{in}. C_{be} is typically from about 10 pF to 50 pF but is dependent on the bias, so the specified value varies (it is larger for larger collector current). In any case, C_{be} is in parallel with the much larger Miller capacitance, $C_{in(Miller)}$ as illustrated in Figure 9-2. For this amplifier you can ignore C_{be}. In a small-signal transistor such as the 2N3904, the base-collector capacitance (C_{bc}) is small (typically 4 pF). The added 100 pF capacitance of C_4 in parallel tends to swamp out the transistor's capacitance and increases the effective C_{bc}. Assume that $C_{bc} = 104$ pF to take into account both the internal capacitance and C_4 in parallel. Then:

$$C_{in} = C_{be} + C_{in(Miller)} \approx C_{in(Miller)}$$

$$\approx C_{bc}(|A_v| + 1) \quad \text{(This is just the Miller capacitance.)}$$

    b. Use the computed gain (unsigned) from Table 9-2 for A_v.

    c. Calculate the equivalent resistance for the input by following the charge/discharge path. In this case, the Thevenin resistance of the source dominates because of its small value. Assume $\beta_{ac} = 170$. The equation for R_{in} is

$$R_{in} = R_{TH} \| R_1 \| R_2 \| \beta_{ac}(r'_e + R_{E1})$$

    d. Calculate the cutoff frequency $f_{c(in)}$ due to the input circuit.

$$f_{c(in)} = 1 / (2\pi R_{in} C_{in})$$

4) In this step, you will compute the critical frequency due to the output network, which consists of $C_{out(Miller)}$, which is in parallel with C_{ce} (the internal capacitance between the collector and emitter). C_{ce} is the least important of the junction capacitances because it is very small, and is much less than $C_{out(Miller)}$. For this reason, you can ignore C_{ce} in this experiment and use $C_{out(Miller)}$ as the output capacitance.

    a. $C_{out} = C_{out(Miller)} = C_{bc}[(A_v + 1) / A_v]$ where A_v is the absolute value of the gain. Enter this as C_{out} in the table.

    b. Calculate $R_c = R_C \| R_L$. Enter this as the output resistance, R_{out}, in the table.

    c. Calculate the cutoff frequency $f_{c(out)}$ due to the input circuit.

$$f_{c(out)} = 1 / (2\pi R_{out} C_{out})$$

5) The calculated overall response will be less than the smaller of the input and output cutoff frequencies. One way to estimate the overall response is:

$$f_{c(overall)} = \frac{1}{\sqrt{\left(\frac{1}{f_{c(in)}}\right)^2 + \left(\frac{1}{f_{c(out)}}\right)^2}}$$

Use this formula to estimate the overall calculated high-frequency cutoff.

6) Measure and record the upper cutoff frequency with the oscilloscope and the Bode plotter, which should be in close agreement. The method was given in Part 1, but the Bode plotter horizontal scale will necessarily be different. Choose a scale that spreads out the horizontal axis in the upper frequency region to make the measurement.

Table 9-5: Amplifier High-Frequency Response Parameters

Parameter	Computed Value	Measured Value
C_{in}		
R_{in}		
$f_{c(in)}$		
C_{out}		
R_{out}		
$f_{c(out)}$		
f_{cu}		

Conclusions for Part 2

Questions for Part 2

1) How does the presence of R_{E1} in the emitter circuit affect both the lower and upper critical frequencies?

2) Why do you think C_{ce} is much smaller than either C_{bc} or C_{be}?

Devices Experiment 10 - Thyristors

10.1 Introduction

Thyristors are semiconductor devices consisting of four alternating layers of *p* and *n* material. Thyristors are primarily used in power control and switching applications. A variety of geometry and gate arrangements are available, leading to various types of thyristors such as the diac, triac, and silicon-controlled rectifier (SCR). In general, SCRs have higher current and voltage ratings, and can handle more power than other thyristors, so they are more widely used in power applications.

In Part 1 of this experiment, you will investigate SCR characteristics. You will start by observing the characteristic curve on an oscilloscope. The SCR is a four-layer device that can be represented as equivalent *pnp* and *npn* transistors. You will test the equivalent transistor circuit and then replace the transistors with an SCR, observing the difference. The circuits are a little different because of the physical differences between the two transistor models and an SCR. You may also find that if you construct the transistor circuit in the lab, a small capacitor between the base of Q_2 and ground will help avoid noise from triggering the circuit; this addition does not work in Multisim, so it has been omitted here.

In Part 2, you will investigate two SCR crowbar circuits, which are widely used as a power supply protection circuit because it can respond very rapidly to an overvoltage. The name crowbar was coined because it was said to be like "putting a crowbar across the output". The crowbar circuit senses the output voltage and responds to an overvoltage situation by shorting the output. If the power supply has current limiting, it will remain on; it is important that power be limited to safe levels. This is commonly accomplished with fold-back current limiting (studied in Chapter 17 of the text), whereby the peak current is limited when the voltage drops below some set value.

10.2 Reading

Floyd, *Electronic Devices, 8th Edition*, Chapter 11

10.3 Key Objectives

Part 1: Measure the gate trigger voltage and holding current for an SCR and test a simulated SCR circuit used for an ac control application.

Part 2: Test two SCR crowbar circuits and report on their performance.

10.4 Multisim Files

Part 1: *Devices_Exp_10_Part_01a, Devices_Exp_10_Part_01b,* and *Devices_Exp_10_Part_01c*

Part 2: *Devices_Exp_10_Part_02a* and *Devices Exp_10_Part_02b*

Part 1: SCR Basics

1) Open the Multisim file *Devices_Exp_10_Part_01a*. The circuit is designed to show the characteristic curve for an SCR with different gate currents. You can see how the forward-breakover voltage ($V_{BR(F)}$) changes as the gate current changes. Figure 10-1 shows the rising trace on the left and the falling trace on the right. If you turn off the circuit at just the right point, you can stop the trace for one or the other to examine it. The frequency has been intentionally set lower to allow you to see both rising and falling traces as they occur.

2) Observe the effect of different gate currents on the forward-blocking region by setting the switches to different positions (using the "A", "B", and "C" keys to control the switches).

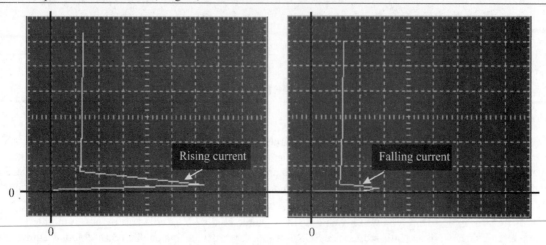

0

0   0

Figure 10-1: SCR Rising and Falling Current Characteristic Curves

Observations:

3)   Open the Multisim file *Devices_Exp_10_Part_01b*. This circuit is a transistor equivalent to an SCR, which is a form of latching circuit. Activate the circuit; the LED should be off. Measure the voltage across the transistors from the emitter of the *pnp* transistor (labeled A) to the emitter of the *npn* transistor (labeled K). This is shown in Table 10-1 as $V_{AK(off\ state)}$. (V_{AK} is the anode-to-cathode voltage.) Enter the measured voltage in Table 10-1 under in the "Transistor Latch" column.

4)   Close J_1 by pressing the "A" key. Slowly decrease the resistance in the gate circuit by pressing the "B" key until the LED comes on. Measure V_{AK} with the LED on (latch closed). Record this value as $V_{AK(on\ state)}$ in Table 10-1. Measure the voltage across R_3. Record this as the gate trigger voltage in Table 10-1.

5)   Open J_1. The LED should stay on because of latching action. The SCR can be turned off by interrupting the anode current or forced commutation. The *RC* circuit consisting of R_5, C_2, and J_2 form a commutation circuit. Close J_2 and observe the result.

6)   Open the Multisim file *Devices_Exp_10_Part_01c*. The transistor latch has been replaced with a 2N6402 SCR. The 2N6402 is one of the 2N6400 series of SCRs whose reverse-blocking voltage increases as the numbers increase. The triggering current is a little larger than that of the transistor latch. Activate the circuit (the LED should be off). Measure $V_{AK(off\ state)}$. Enter the measured voltage in Table 10-1 in the "SCR" column.

7)   Close J_1 by pressing the "A" key. Slowly decrease the resistance in the gate circuit by pressing the "B" key until the LED comes on. Measure V_{AK} with the LED on (latch closed). Record this value as $V_{AK(on\ state)}$ in Table 10-1. Measure the voltage across R_3. Record this as the gate trigger voltage in Table 10-1.

Table 10-1: SCR Parameters

Parameter	Transistor Latch	SCR
$V_{AK(off\ state)}$		
$V_{AK(on\ state)}$		
$V_{Gate\ trigger}$		

Observations

Conclusions for Part 1

Questions for Part 1

1) How would you calibrate the oscilloscope in step 1 for current in the vertical axis?

2) Explain how capacitor commutation works in the SCR circuit.

Part 2:   SCR Crowbar Circuits

1) Open the Multisim file *Devices Exp_10_Part_02a*. This circuit is a basic crowbar circuit that causes the SCR to short the output of a power supply. This circuit is designed to protect a 5 V circuit from an overvoltage from the power supply. In order to make a variable voltage in Multisim, notice that a small potentiometer is placed across the source. This combination represents the variable power supply. While this is not particularly practical, it allows you to actively control the supply voltage by using the "A" key and the "Shift-A" combination. Investigate this circuit. Discuss your findings and explain how the circuit works, including a discussion of the SCR triggering.

    Observations:

2) Open the Multisim file *Devices Exp_10_part_02b*. This circuit is a modification of the circuit in step 1 and can sense an overvoltage condition and use it to close (or open) relay contacts that can initiate some action. In this case, the circuit is not strictly a crowbar circuit as drawn but can easily be modified to turn off power completely by using a separate mechanically latching relay. In this circuit, there is no limiting of the output voltage. Notice that the circuit uses a different zener diode to keep the trigger voltage about the same as in step 1. Investigate this circuit. Discuss your findings and explain how the circuit works.

    Observations:

Conclusions for Part 2

Questions for Part 2

1) Without changing the zener diode, could you change the circuit in step 1 so that you could adjust the crowbar voltage by a small amount to compensate for differences in zener diodes? Explain your answer.

2) What modification would you make to the circuit in step 2 that would cause the voltage to drop immediately after an overvoltage condition was sensed?

Devices Experiment 11 - The Operational Amplifier

11.1 Introduction

Most op-amps have a differential amplifier as the input stage. The differential amplifier is an important circuit that you should understand, so this experiment begins with a discrete differential amplifier. There are important advantages to differential amplifiers, particularly for noise rejection, which is why differential amplifiers are used for the input stage. By testing a discrete differential amplifier, you will learn about the inner workings of integrated circuit op-amps. Unlike earlier experiments with discrete circuits, you will need only to calculate a minimum of dc parameters, but the focus is on signals.

In Part 1 of this experiment, you will compare various ways a signal can be delivered to or extracted from the amplifier. You will compare the results of a differential input with a common-mode input and then test the circuit with both inputs applied together. You may be surprised at the results! You will also measure the CMRR for the amplifier. Recall that the CMRR is the ratio of differential gain to common-mode gain.

In Part 2, you will be introduced to op-amps. Before delving into basic op-amp circuits, you will look at ways to measure some important parameters for op-amps. Whenever you measure parameters for any electronic device, it is a good idea to include a schematic of the specific test circuit and conditions for the test. This enables another person to repeat exactly what you have done. For this experiment, you will measure the input offset voltage, input bias current, input offset current, CMRR, and slew rate. Schematics of the setup in the lab are included.

The methods you will use for testing are applicable to various op-amps, but the LM741 is selected because it is a reasonable choice for many basic circuits and is inexpensive. Although newer designs have greatly improved specifications, the LM741 is still an industry standard more than 40 years after its introduction. It is a general-purpose, high-gain amplifier that is widely used in school electronics labs.

11.2 Reading

Floyd, *Electronic Devices, 8th Edition*, Chapter 12 (review Section 6-7)

11.3 Key Objectives

Part 1: Calculate and measure the differential and common-mode gains for a differential amplifier and observe the amplifier with both a differential and common-mode signal at the same time.

Part 2: Measure basic parameters for a simulated LM741 op-amp.

11.4 Multisim Files

Part 1: *Devices_Exp_11_Part_01a* amd *Devices_Exp_11_Part_01b*

Part 2: *Devices_Exp_11_Part_02a* through *Devices_Exp_11_Part_02d*

Part 1: Operational Amplifier Characteristics

11.5 The Differential Amplifier

1) Open the Multisim file *Devices_Exp_11_Part_01a*. The circuit is initially set up with a single-ended differential input and single-ended output; the output will be taken from the collector of Q_2. To simplify the calculations, the dc equation for I_E is set up to allow you to find r'_e. You can then use this resistance to quickly calculate the gain. The circuit works best if the transistors, the base resistors and the emitter resistors are matched (which, of course, they are in Multisim). Given this condition, the equation for I_E can be developed by writing Kirchhoff's voltage law around the base circuit of Q_1 (and assuming I_E for each transistor is identical; they should be close).

$$-I_{B1}R_{B1} + V_{BE} + I_E R_{E1} + 2I_E R_T - V_{EE} = 0$$

$$I_E = \frac{V_{EE} - V_{BE}}{\dfrac{R_{B1}}{\beta_{DC}} + R_{E1} + 2R_T}$$

a. Use the equation given to solve for I_E. For V_{EE}, the sign is taken into account in writing the loop, so use the absolute (unsigned) value. Assume $\beta_{DC} = 170$. Enter the calculated value in Table 11-1.

b. Calculate r'_e from $r'_e = I_E / 25$ mV. Enter the calculated value in Table 11-1.

c. Calculate the differential voltage gain from $A_{v(d)} = R_C / [2(r'_e + R_E)]$. This equation is equivalent to the gain equation for a CB amplifier except for the 2 in the denominator. The reduction by a factor of 2 is due to the attenuation of the signal to point A (shown in the Multisim file) by Q_1, which looks like a CC amplifier to the input signal with a gain of 0.5 to point A.

d. Calculate the peak output voltage, assuming the input voltage is 35 mV$_p$.

2) Measure the quantities listed in Table 11-1 (except r'_e). To measure I_E, zero the ac source, measure the voltage across R_T, and divide by $2R_T$; this has the effect of averaging any difference in the two emitter currents. To measure the gain, reset the input signal to 35 mV$_p$ and measure V_{in} and V_{out} with the oscilloscope and calculate the ratio of V_{out}/V_{in}. To view both the input and the single-ended output voltage, connect Channel 1 to the source and Channel 2 to the collector of Q_2. Couple the signals in each channel using **AC Coupling** (Go to the **CH 1 MENU** to select the coupling).

Table 11-1: Computed and Measured Differential Amplifier Values

Quantity	Computed Value	Measured Value
I_E		
r'_e		
$A_{v(d)}$		
$V_{out} = V_c$		

3) In this step, you will observe a double-ended output from this circuit. To observe the output as a double-ended signal (also called a differential output), do the following:

a. Move the Channel 1 probe to the collector of Q_1.

b. Set both channels to "1 V/Div".

c. From the **MATH MENU**, select **CH1 – CH2**.

d. Activate the circuit. You should see three signals: one from CH1 (on Q_1), one from CH2 (on Q_2), and a third signal representing the difference between CH1 and CH2. The third signal is the differential output, which is not referenced to ground. Most scopes provide this method of viewing a differential signal. In addition, many higher-end scopes offer special differential probes (not available in Multisim) that are optimized for this purpose. Differential probes are particularly useful when time differences in the two signals are critical.

Observations:

4) In this step, you will change the differential input from single-ended (referenced to ground) to double-ended (one that is not referenced to ground). The steps to change the circuit are:

a. Deactivate the circuit.

b. Delete the ground connection from the source and from C_2.

c. Connect a wire from the bottom of the source to the bottom of C_2. This is a common arrangement for connecting a remote sensor to a circuit.

d. Activate the circuit.

Observations:

5) Open the Multisim file *Devices_Exp_11_Part_01b*. The circuit is the same differential amplifier except a simulated 4900 Hz noise signal is added (named CM$_{noise}$). The simulated noise is applied in *common-mode*, meaning that it affects both sides of the amplifier equally. This happens if the wires connecting a sensor (for example) intercepts noise. The noise will affect both wires equally, producing a common mode inference signal. Notice that the common-mode signal is approximately 3X larger than the original differential signal. Calculate and record the quantities shown in Table 11-2.

    a.   The single-ended common-mode gain, A_{cm}, is calculated from:

$$A_{cm} = R_C / (r'_e + R_{E1} + 2R_T)$$

    b.   The single-ended common-mode output voltage is the input voltage multiplied by the common-mode gain.

6) In the Multisim file, the differential input signal is initially set to zero and the scope is set up to look at the single-ended common-mode signals on each channel as well as the differential common-mode signal. There are three signals on the scope: the two single-ended common-mode outputs and the differential output. Activate the circuit and measure the single-ended common-mode signal at either Q_1 or Q_2 (they should be the same; in fact you may not realize at first that there are two overlapping signals on the screen but you can separate them with the **POSITION** controls). Record this as the measured V_{out} in Table 11-2 and use this value to calculate the measured A_{cm} for the single-ended common mode gain. Summarize your observations.

Table 11-2: Differential Amplifier Common Mode Calculations and Measurements

Quantity	Computed Value	Measured Value
A_{cm}		
$V_{out} = V_c$		

7) Deactivate the circuit and set the differential input to 35 mV$_p$. This represents a small differential signal (the 1 kHz signal) in the presence of the much larger common-mode simulated noise (the 4900 Hz signal). Frequently, the common-mode signal is a lower frequency (such as 60 Hz power line interference), but the higher frequency makes it easier to detect the difference between the differential and the common-mode signals, hence it was used in this experiment. Observe the single-ended output signals at each collector and the differential output signal. Summarize your observations:

Conclusions for Part 1

Questions for Part 1

1) What is the difference between differential gain and common-mode gain?

2) For the single-ended output signal, what is the CMRR for the amplifier in this experiment? Use your measured values to calculate CMRR.

Part 2:   Op-Amp Specifications and Basic Parameters

1) Table 11-3 lists several important specifications for the 741C op-amp for $T_A = 25°C$. You may want to confirm these specifications on the manufacturer's specification sheet, which is available at http://www.fairchildsemi.com/ds/LM/LM741.pdf.

Table 11-3: 741 Op-Amp Specifications

Step	Parameter	Specified Value			Measured Value
		Minimum	**Typical**	**Maximum**	
2b	Input offset voltage, V_{OS}		2.0 mV	6.0 mV	
3b	Input bias current, I_{BIAS}		80 nA	500 nA	
3c	Input offset current, I_{OS}		20 nA	200 nA	
4a	Differential gain, $A_{v(d)}$				
4b	Common-mode gain, A_{cm}				
4c	CMRR′	70 dB	90 dB		
5	Slew rate		0.5 V/μs		

2) Open the Multisim file *Devices_Exp_11_Part_02a*. The circuit will enable you to measure the input offset voltage, V_{OS}, of the 741C op-amp. The input offset voltage is the amount of voltage that must be applied between the input terminals through two equal resistors to give zero output voltage. It is a dc parameter. The LM741 is shown in the Multisim circuit with decoupling capacitors connected to the power supplies. For Multisim, these capacitors have no effect on the results, but are included for good practice.

    a. Connect the DMM and measure the output voltage, V_{OUT}, with no input signal (ideally the output is 0 V). The input offset voltage is found by dividing the output voltage by the closed-loop gain of the amplifier, treating it as if it were a noninverting amplifier. In this case, the closed-loop gain is found from:

$$A_{cl(NI)} = 1 + R_f / R_i.$$

    b. Record the measured V_{OS} in the first line in Table 11-3.

3) In this step, you will measure the input bias current, I_{BIAS}, and the input offset current, I_{OS}. Open the Multisim file *Devices_Exp_11-part_2b*. The input bias current is the average of the input currents at each input terminal. The input offset current is a measure of how well these two currents match. The input offset current is the difference in the two bias currents. The input bias current and input offset current are dc parameters. Steps for the measurements are:

    a. Measure the voltage across each resistor and apply Ohm's law to calculate the current in each resistor. In Multisim, you could insert ammeters in series with the inputs, but they are generally not practical for making sensitive measurements like this in actual lab situations.

  b. Record the *average* of these two currents in Table 11-3 as the input bias current, I_{BIAS}.

  c. Record the *difference* in these two currents in Table 11-3 as the input offset current, I_{OS}.

4) Open the Multisim file *Devices_Exp_11_Part_02c*. This circuit is set up to measure the common-mode rejection ratio, CMRR'. The CMRR' is 20 times the logarithmic ratio of the op-amp's differential gain ($A_{v(d)}$) divided by the common-mode gain (A_{cm}). Because it is a ratio of gains, CMRR' is an ac parameter. It is frequently expressed in decibels (and shown with the prime symbol to indicate this). The definition for CMRR' is

  $\text{CMRR}' = 20 \log [A_{v(d)} / A_{cm}]$

  a. Place both switches in the differential-mode position (the space bar moves both switches together). Connect the oscilloscope so that one channel is on the output and the other one is on the differential signal source. Measure differential gain by dividing the output voltage by the differential input voltage. Enter the measured differential gain, $A_{v(d)}$, in Table 11-3.

  b. Place both switches in the common-mode position. Connect the oscilloscope so that one channel is on the output and the other one is on the common-mode signal source. Measure common-mode gain by dividing the output voltage by the differential input voltage. (Notice that a much larger input signal is used for the common mode gain measurement because common-mode gain is small. Enter the measured common-mode gain, A_{cm}, in Table 11-3.

  c. Determine the measured value of CMRR' (in decibels) for your LM741C op-amp. Record the result in Table 11-3.

5) In this step, you will measure the slew rate for the LM741C. Open the Multisim file *Devices_Exp_11_Part_02d*. The circuit is a unity-gain amplifier with a large-amplitude step function applied to the input. Slew rate is a limitation that affects large signals more than small signals. It is usually specified for a unity-gain voltage-follower with a fast rising input pulse. Slew rate is usually expressed in units of volts/microsecond (V/μs).

Observe the input and output signals with the oscilloscope. The output voltage will be slew-rate limited and will not respond instantaneously to the change in the input voltage. The slew rate can be measured by observing the change in voltage divided by the change in time at any two points on the rising output waveform as shown in Figure 11-1. You might want to use the cursors on the oscilloscope to simplify the measurement. Record the measured value in Table 11-3.

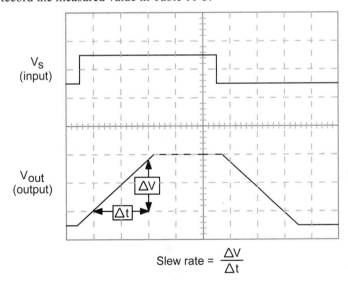

Figure 11-1: Measurement of Slew Rate

Conclusions for Part 2

Questions for Part 2

1) How is a common-mode signal applied to an op-amp?

2) What is the difference between the input bias current and the input offset current?

3) What is the advantage of a high value of CMRR'?

Devices Experiment 12 - Basic Op-Amp Circuits

12.1 Introduction

The op-amp is one of the most versatile electronic components. It can be configured as a comparator, amplifier, voltage converter, summing amplifier, integrator, differentiator, filter, oscillator and many other circuits. In this experiment, you will investigate some of these basic circuits in some detail. You will start with the comparator.

A comparator senses a voltage difference between two inputs and outputs a voltage that is near the positive or negative power supply voltage to indicate which input is larger. To make the comparator circuit sensitive and respond quickly, it needs to have a high open-loop gain and a fast slew rate, which requires low input capacitance and low bias current. Comparators are often used to compare an input with a reference voltage. Special integrated circuits are designed to optimize these requirements and often have other features, like an internal voltage reference, built-in hysteresis, or the ability to set up specific output voltages. Part 1 of this experiment introduces two comparators, which are both useful for interfacing to logic circuits. You will start with an LM160H complementary output comparator that can provide TTL-compatible outputs. Complementary outputs mean that when one output is high, the other is low and vice-versa. The complementary outputs will be used to construct a comparator with positive and negative thresholds, as you will see. Then a MAX941CPA comparator is introduced. It also has a TTL-compatible output and has the ability to put the outputs in a high impedance state, a useful feature for interfacing to a bus.

In Part 2, you will test a basic inverting amplifier and some variations of the inverting amplifier, including a summing amplifier, an integrator, and a differentiator. After observing the waveforms, you will summarize your observations. Look for a common thread in these circuits. At the heart of each is a basic inverting amplifier.

12.2 Reading

Floyd, *Electronic Devices, 8th Edition*, Chapter 13

12.3 Key Objectives

Part 1:   Measure the input and output signals for various comparator circuits and plot the hysteresis curve for a Schmitt trigger circuit.

Part 2:   Observe several circuits using op-amps; measure the responses and describe the results.

12.4 Multisim Files

Part 1:   *Devices_Exp_12_Part_01a*, *Devices_Exp_12_Part_01b*, and *Devices_Exp_12_Part_01c*

Part 2:   *Devices_Exp_12_Part_02a* through *Devices_Exp_12_Part_02d*

Part 1:   Comparator Circuits

1)   Open the Multisim file *Devices_Exp_12_Part_01a*. Connect the Tektronix oscilloscope so that channel 1 is connected to the input signal (marked V_{in}) for the comparator and channel 2 is connected to the non-inverting output ($V_{out\,(+)}$).

2)   Turn on the oscilloscope and activate the circuit. Adjust the scope to show two cycles of the input sine wave.

3)   With the potentiometer set to 50%, sketch the input and output waveforms in Figure 12-1. Show the volts per division settings and the 0 V level for the input and the output.

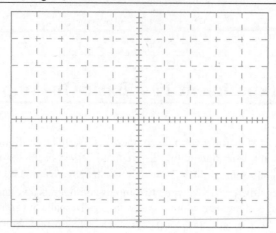

Figure 12-1: Input and Output Waveforms for the LM160H Comparator

4) Vary the potentiometer and describe your observations:

5) Open the Multisim file *Devices_Exp_12_Part_01b*. The circuit has been modified to delete the potentiometer but now has hysteresis. Connect the Tektronix oscilloscope so that channel 1 is connected to the input signal (V_{in}) and channel 2 is connected to the non-inverting output ($V_{out\,(+)}$).

6) Turn on the oscilloscope and activate the circuit. Adjust the scope to show two cycles of the input sine wave. Notice that the rise and fall trigger points are no longer at the 50% level.

7) Sketch the input and output waveforms in Figure 12-2. Show the volts per division settings and the 0 V level for the input and the output.

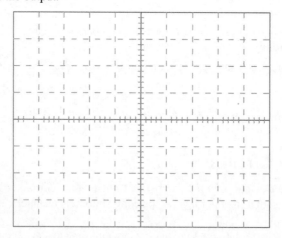

Figure 12-2: Input and Output Waveforms for the LM160H Comparator with Hysteresis

Observations:

8) Open the Multisim file *Devices_Exp_12_Part_01c*. This circuit is another way of achieving hysteresis. This circuit uses positive feedback and is commonly called a Schmitt trigger. An interesting application for the Schmitt trigger is to combine it with an op-amp integrating circuit to form a triangle wave generator (see Experiment 15). The IC is a low-voltage MAX941CPA, with very low bias current and has a single TTL compatible output. In addition to the inverting and noninverting inputs, the MAX941CPA has two additional active-LOW inputs (that is, they take effect, or are active, when LOW). One is SHDN', which shuts down the IC. The other is LATCH', which holds the output state. (Note that the apostrophe after a signal name denotes an active-LOW signal.)

9) Set up the oscilloscope to view the input on Channel 1 and output on Channel 2. Activate the circuit and adjust the oscilloscope to show two cycles on the display. To smooth variations in the output signal, press the **ACQUIRE** button and select **Average**. Notice that the input signal is configured to go slightly below ground. According to the manufacturer, the input can go to 0.3 V below ground. Sketch the input and output waveforms in Figure 12-3.

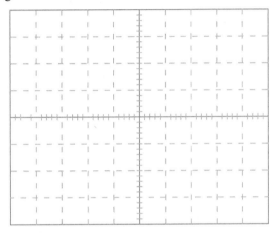

Figure 12-3: Input and Output Waveforms for the MAX941CPA Comparator with Hysteresis

10) Go into the **DISPLAY** menu and choose **Format XY**. This converts the display to show the output voltage plotted as a function of the input voltage (this is called the *hysteresis curve* or *transfer curve*). Sketch the waveform you see in Plot 12-4 and label your plot. To understand the display, keep in mind that the input is plotted on the *x*-axis and the output is plotted on the *y*-axis. Starting at the top left, the input in increasing as it moves to the right. When the input reaches a certain voltage, it drops (vertical line on the right side), which is the upper switching threshold. Think about how the rest of the curve is formed on the scope as you view the signal.

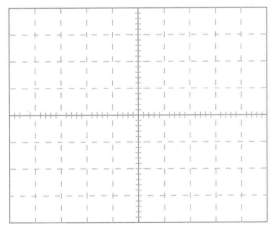

Figure 12-4: Hysteresis Curve for the Schmitt Trigger Circuit

11) Try changing the value of R_2 and observe the effect on the curve.

Observations:

Conclusions for Part 1

Questions for Part 1

1) What is the purpose of the voltage divider on the input of the comparator circuits?

2) The output duty cycle for the comparator circuit in *Devices_Exp_12_Part_01a* could be varied from 0% to 100%. How would you specifically modify the circuit to restrict the duty cycle to a range between 10% and 90% for the input signal given?

3) Reading the graph in Figure 12-4, what are the two threshold voltages for the Schmitt trigger?

Part 2: Op-Amp Circuits

12.5 Inverting Amplifier

1) Open the Multisim file *Devices_Exp_12_Part_02a*. This circuit is a basic inverting amplifier using an LM741. Calculate the gain from the equation $A_{cl(I)} = -(R_f / R_i)$. The negative sign is shown to account for the inversion. Record the computed value in Table 12-1.

2) Calculate V_{out} using the computed closed-loop gain. Record it as the computed value in Table 12-1.

3) Activate the circuit and use the Tektronix oscilloscope to measure the parameters listed in Table 12-1. Determine the gain by dividing the output voltage by the input voltage.

Table 12-1: Computed and Measured Values for Inverting Amplifier

Parameter	Computed Value	Measured Value
V_{in}	500 mV$_{pp}$	
$A_{cl(I)}$		
V_{out}		

12.6 Summing Amplifier

1) Open the Multisim file *Devices_Exp_12_Part_02b*. This circuit is a modification of the previous circuit but illustrates a 5-input summing amplifier. Although this is not a practical circuit for various reasons, it does illustrate an important idea in electronics. Rather than explain the circuit in detail, this one is for you to

observe and describe what it does (and why!). To help, the scope is already set up. Activate the circuit, and then close the switches, one at a time from the top. Summarize your observations and what is happening.

Observations:

2) Open the Multisim file *Devices_Exp_12_Part_02c*. This circuit illustrates another summing amplifier application. To simplify setup, the scope is already connected. Activate the circuit and describe how it works. Notice that the trigger channel for the scope is Channel 3. Can you figure out why this is the best choice? (If you aren't sure, use the settings in **TRIGGER MENU** to try different trigger channels.)

Observations:

12.7 Differentiating and Integrating Circuits

1) Open the Multisim file *Devices_Exp_12_Part_02d*. This circuit is driven by a square wave from the function generator. Try figuring out what V_{out1} should look like before activating the circuit. Then activate the circuit and observe the waveforms on the oscilloscope. Try changing the value of C_f a little up or down and observe the effect on the output. Restore it to 10 nF for the final step.

Observations:

2) Deactivate the circuit and connect Channel 3 to V_{out2}. Predict what will happen when the circuit is activated. Then activate the circuit and observe the result.

Observations:

Conclusions for Part 2

Questions for Part 2

1) Assume you had to connect three microphones to a single amplifier. You want to be able to turn on or off each microphone independently and control the gain within a limited range of each microphone independently. Describe the circuit you would use to accomplish this.

2) What evidence did you see that differentiation is the opposite of integration?

Devices Experiment 13 - Special-Purpose Op-Amps

13.1 Introduction

There are many applications that require a special-purpose op-amp, which optimizes certain op-amp characteristics. One of these is the instrumentation amplifier (IA), which is useful where a small signal is in a noisy environment (such as biomedical sensors, industrial sensors, microphones, and so forth). In these cases, the CMRR' is one of the most important specifications. IAs are available in a single IC package, optimized for these applications.

In Part 1 of this experiment, you will test an IA. A 4 kHz source will simulate a low-level (20 mV) signal differential-input signal from a transducer. A 1.5 kHz signal will simulate a large noise source, so it is connected in common-mode, and you will be able to see how effective the IA is for rejecting this simulated noise while passing the desired signal. When the circuit is properly adjusted, the oscilloscope will not show the noise on the output, but the noise can still be detected with an instrument called the spectrum analyzer. You will use the spectrum analyzer to determine the amplitude of the noise on the output and the CMRR'. After testing an IA constructed from 741s, you will test a LT1101 instrumentation amplifier in a typical remote sensing application. This IC has some interesting capabilities. It is the first IA that can be powered from a single supply voltage and draws as little as 75 μA of current, making it ideal for battery-powered applications. It also does not require an external gain selection resistor and has two selectable internal gains (10 and 100). It has excellent specifications. For example, the CMRR' is typically 100 dB and the input offset current is only 130 pA.

In Part 2, you will investigate another special-purpose op-amp: the operational transconductance amplifier (OTA). The OTA is a versatile IC that uses current-control in applications ranging from current-controlled amplifiers, impedances, filters, oscillators and more. (A good resource for these applications is the manufacturer's data sheet – available at http://www.national/com/ds/LM/LM13700.pdf). You will test a current-controlled amplifier and plot the gain as a function of bias current for this amplifier. Following your investigation of this circuit, you will test a log/antilog combination constructed with virtual op-amps. This circuit can be constructed as shown in the Multisim file with three LM741 op-amps (and it works well). In the early days of computers, log amps were employed to do various mathematical operations (hence the name operational amplifier). Today, they find more application in high-frequency signal compression and detection, and there are various log/antilog devices especially designed for these uses.

13.2 Reading

Floyd, *Electronic Devices, 8th Edition*, Chapter 14

13.3 Key Objectives

Part 1:   Measure parameters for an instrumentation amplifier constructed from op-amps, including CMRR'. Measure parameters for a simulated measurement system using an L1101 instrumentation amplifier.

Part 2:   Test a current-controlled amplifier and plot the gain as a function of bias current. Test a basic log/antilog circuit designed to take square roots.

13.4 Multisim Files

Part 1:   *Devices_Exp_13_Part_01a* and *Devices_Exp_13_Part_01b*

Part 2:   *Devices_Exp_13_Part_02a* and *Devices_Exp_13_Part_02b*

Part 1:   The Instrumentation Amplifier

13.5 The Discrete Instrumentation Amplifier

1)   Open the Multisim file *Devices_Exp_13_Part_01a*. The circuit has two sources – a 20 mV$_p$, 4 kHz signal that is applied between the differential inputs, and a 1.0 V$_p$, 1.5 kHz common-mode simulated noise signal.

These frequencies are unrelated to each other and were selected so that you can clearly distinguish between the differential and common-mode signals. The Tektronix oscilloscope is already set up to view both signals.

2) From the resistor values in the circuit, calculate the expected differential gain for the instrumentation amplifier. The gain equation is $A_{v(d)} = 1 + (2R_1 / R_G)$. It is shown here as $A_{v(d)}$ to emphasize that it is a differential gain. Enter the computed gain in Table 13-1.

3) Using the computed gain, calculate the expected differential output voltage. Enter the expected output voltage in Table 13-1.

4) Activate the circuit and turn on the oscilloscope. Notice that one of the signals is not stable and tends to drift across the screen. This is because the two sources are not integer multiples of each other, so the oscilloscope cannot synchronize both signals. (Some oscilloscopes can allow for a separate trigger for each channel and thus avoid this problem.) You can "freeze" either one of the signals for display by changing the trigger source (either CH1 or CH2).

5) Measure $V_{out(p)}$. You may find it easier to get an accurate reading using the cursors and dividing the peak-to-peak reading by 2. Notice that there is a slight variation in the output differential signal because the common-mode signal is not entirely cancelled at this point. Enter your measured value in Table 13-1.

6) Based on the measured input and output voltages, calculate a measured gain. Record the result in Table 13-1. Deactivate the circuit for the next step.

Table 13-1: Calculated and Measured Discrete IA Values

Parameter	Calculated Value	Measured Value
$A_{v(d)}$		
$V_{out(p)}$		

7) In this step, you will connect the spectrum analyzer to the circuit and set it up to see the spectrum. You will see that it is actually easier to obtain the best common-mode rejection by making adjustments while observing the spectrum. Connect the spectrum analyzer so that the output of the circuit is connected to the **IN** terminal. Then adjust the controls as follows:

   a. In the **Span Control** section, choose **Set Span**.

   b. In the **Frequency** section, you can choose certain options and allow the others to be set by Multisim:

      i. In the **Span** window, do not enter a value. Instead, you can select the start and stop frequencies.

      ii. In the **Start** window, enter "1 kHz". This will be the frequency on the left side of the display.

      iii. In the **Center** window, do not enter a value; Multisim will calculate a center frequency.

      iv. In the **End** window, enter "9 kHz". This will be the frequency at the right side of the display; it is high enough to allow you to see the first harmonic and any nearby interactions.

      v. Click the **Enter** button to put these values into the spectrum analyzer memory.

   c. In the **Amplitude** section, choose the following:

      i. Select **dB** to set up the vertical scale as a log scale. This will enable you to see tiny signals that are not easily seen with a linear display.

      ii. In the **Range** window, choose "20 dB/Div". This will enable you to see both the large differential signal and the extremely small common-mode signal on the display together.

      iii. In the **Ref** window, choose "0 dB".

      iv. In the **Resolution Freq.** windows, enter "100" in the left window with units of "Hz" in the right window.

d. Click the **Set...** button on the lower right side of the instrument. This will open the **Settings** menu. Click **FFT Points** and select "32768" from the drop-down list to show the maximum number of points, Click **Accept**.

8) Activate the circuit. The spectrum analyzer will take a few cycles to respond, so wait until it settles. The highest peak is the differential signal, and the second highest peak is the common-mode (CM) signal. You can confirm this by reading the frequency when the cursor is over each peak. To minimize the CM signal, use the slider near the potentiometer, or press the "A" or "Shift + A" keys, to adjust R_{6B}. You should adjust the potentiometer for the minimum CM signal and leave it in that position for the next step. After adjusting it, notice what happens to the oscilloscope view of the output signal. Summarize your observations.

9) On the spectrum analyzer, read the amplitude of the common-mode signal and the amplitude of the differential signal in decibels. Follow these steps:

a. To measure a point accurately, you must right-click the small triangle in the upper right corner of the spectrum analyzer and choose **Set X_Value**. This will open up a menu item that allows you to input the desired frequency that you wish to measure. Type "4k" into the box and click the **OK** button. The spectrum analyzer will perform the required interpolation and display 4.000 kHz and the decibel level referenced to 1.0 V_p. Record the decibel level in Table 13-2.

b. Convert the dB reading to a voltage reading by solving for $V_{out(p)}$ from

$$dB = 20 \log [V_{out(p)} / 1.0 \ V_p].$$

Enter the voltage reading in Table 13-2. You should have excellent agreement between the value read on the spectrum analyzer and the oscilloscope.

c. Repeat steps a and b for the 1.5 kHz common-mode signal. The common-mode output is shown in the table as $V_{out(p)(cm)}$.

d. Determine the common-mode gain, A_{cm}, by dividing the common-mode output by the common-mode input (1 V_p). Record the common-mode gain in Table 13-2.

10) Use the differential gain $A_{v(d)}$ from Table 13-1 and the common-mode gain $A_{v(cm)}$ from Table 13-2 to verify that CMRR' = 10 log $[A_{v(d)} / A_{cm}]$ is over 100 dB. Record CMRR' in Table 13-2.

Observations:

Table 13-2: Measured Spectrum Analyzer Values

Parameter	Measured Value
V_{out} (dB) @ 4 kHz	
$V_{out(p)}$ (V_p) @ 4 kHz	
$V_{out(p)(cm)}$ (dB) @ 1.5 kHz	
$V_{out(p)(cm)}$ (V_p) @ 1.5 kHz	
A_{cm}	
CMRR'	

13.6 The LT1101 Instrumentation Amplifier

1) Open the Multisim file *Devices_Exp_13_Part_01b*. This circuit uses an integrated circuit IA that is suited for remote measurements such as those simulated in this circuit. R_1 represents a sensitive resistive transducer such as a strain gauge; it is split into two resistors (R_{1A} and R_{1B}) to give you more control of the small resistance change that is typical of many resistive transducers. Notice that each 1% change of R_{1A} represents only 10 mΩ of resistance change. The LT1101 can operate on much smaller power supply voltages for very low-power applications, but it is more sensitive to change with the higher voltage used here. The common-mode signal simulates how a capacitively coupled noise source might affect the signal path; the CM signal is monitored by the scope.

2) Activate the circuit.

3) Increase R_{1A} in 10 mΩ (1%) increments by pressing the "A" key.

4) Monitor the dc output with the DMM. Record the output voltage, using three significant digits for each setting in Table 13-3.

5) Plot the data in Plot 13-1 and summarize your observations. As part of your summary, include a measurement of the common-mode output. This time you can use the DMM to measure the common-mode gain directly. This works because there is only one ac signal to measure (the common-mode signal). Select the **dB** scale and ac mode. Click the **Set...** button; this brings up the **Multimeter Settings** menu. Change the **dB Relative Value (V)** to "70.7" and "mV" to set the reference level to the input rms value. This makes the reading of the output relative to the input, so you are directly reading $A_{v(d)}$. You can easily find the CMRR' because $A_{v(d)}$ is set by the manufacturer to be 100.

Table 13-3: LT1101 Measurements

R_{1A} Setting	V_{OUT}	R_{1A} Setting	V_{OUT}
330.00 Ω		330.06 Ω	
330.01 Ω		330.07 Ω	
330.02 Ω		330.08 Ω	
330.03 Ω		330.09 Ω	
330.04 Ω		330.10 Ω	
330.05 Ω			

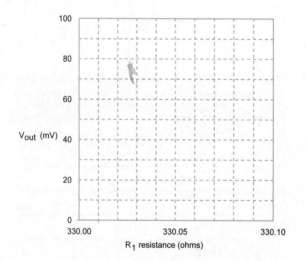

Plot 13-1: LT1101 *V-R* Measurements

Conclusions for Part 1

Questions for Part 1

1) What are major advantages for measuring the signals in *Devices_Exp_13_Part_01a* with a spectrum analyzer?

2) Based on your data, what is the CMRR' for the circuit using the LT1101?

Part 2:   The LM13700 Operational Transconductance Amplifier

13.7   Current-Controlled Audio Amplifier

Open the Multisim file *Devices_Exp_13_Part_02a*. The circuit is a current-controlled audio amplifier using one-half of an LM13700 IC. The Multisim file uses a potentiometer as a simple way to vary the bias current (rather than resetting a current source each time). The LM13700 has two OTAs, each with differential inputs and a push-pull output. In addition, optional buffer transistors can be connected to the output. Connect the oscilloscope so that channel 1 monitors the input and channel 2 monitors the output. Activate the circuit and adjust the scope to view the input and output.

1) The input voltage is set to 100 mV$_p$. Set the gain to maximum by moving R_{gain} to 0%, which supplies maximum bias current (approximately 2.0 mA).

2) Record the measured bias in Table 13-4. A potentiometer may seem contrary to the idea of voltage-control, but is simpler to adjust in Multisim, so the circuit uses it in place of a variable voltage source.

3) Using the oscilloscope, measure the peak output voltage ($V_{out(p)}$) and record it in Table 13-4. Use $V_{out(p)}$ and $V_{in(p)}$ to calculate the voltage gain, A_v.

4) Repeat steps 1 through 3 for each setting of R_{gain} shown in Table 13-4.

Table 13-4: Current-Controlled Audio Amplifier Measured and Calculated Values

R_{gain} Setting	I_{BIAS}	$V_{out(p)}$	Voltage Gain, A_v
0%			
20%			
40%			
60%			
80%			
100%			

5) Plot gain A_v as a function of I_{BIAS} in Plot 13-2.

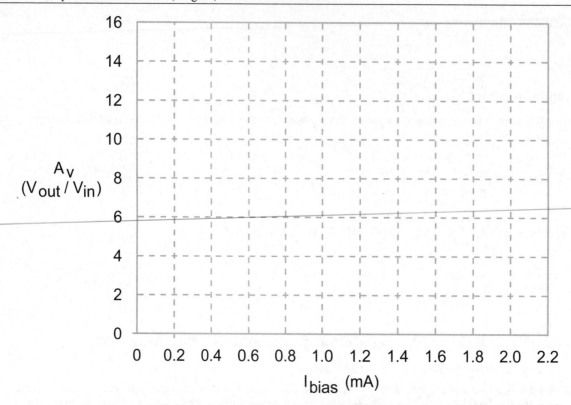

Plot 13-2: Current-Controlled Audio Amplifier Gain Plot

Summarize your observations.

13.8 A Log/Antilog Combination

1) Open the Multisim file *Devices_Exp_13_Part_02b*. The circuit is an application of a log/antilog amplifier. Virtual op-amps are shown for simplicity, but the circuit has been constructed in lab with the old reliable LM741C (but was limited to 12 V maximum input). The circuit takes the square root of the input voltage by dividing the log value by 2 and taking the antilog, as

$$\sqrt{X} = X^{1/2} = \text{antilog}\,[\log\,(X^{1/2})] = \text{antilog}\,[1/2\,\log\,(X)]$$

You should be able to follow the circuit and see that it is a square root circuit! The output voltage is the square root of the input voltage. Before the circuit can determine the square root accurately, it must be calibrated.

2) Complete the middle (predicted) output voltage by calculating the square root of the V_{IN} for each value of V_{IN} and recording it in Table 13-5.

3) Set the input voltage to 4.0 V by double-clicking the source (V_{IN}); then activate the circuit.

4) Adjust the calibrate potentiometer so that the output is as close as possible to 2.00 V. The increment value in Multisim is set for 0.1% to enable you to set a precise value. Enter the measured V_{OUT} in Table 13-5 for the 4 V input.

5) Repeat step 4 for each value in Table 13-5. You must stop the simulation, double-click V_{IN} to set a new value, and restart the simulation to do so. Summarize your observations.

Table 13-5: Square Root Circuit Predicted and Measured Values

V_{IN} Setting	V_{OUT} (predicted)	V_{OUT} (measured)
1 V		
2 V		
4 V		
8 V		
12 V		
16 V		

Conclusions for Part 2

Questions for Part 2

1) Based on your data, what bias current is required to produce a gain of 10 with the OTA circuit?

2) How would you change the square root circuit to a cube root circuit?

Name _____   Class _____

Date _____   Instructor_____

Devices Experiment 14 - Active Filters

14.1 Introduction

A filter is a circuit that produces a prescribed frequency response. Active filters contain resistors, capacitors, and sometimes inductors as well as an active element such as an operational amplifier. Active filters can achieve frequency response characteristics that are nearly ideal at a reasonable cost for frequencies up to about 100 kHz. Above this, active filters are limited by bandwidth.

The Butterworth filter, which is one type of filter, has the flattest passband characteristic. Since a flat passband is generally the most important filter characteristic, it will be investigated in Part 1 of this experiment. You will calculate three points on the response curve of a 4-pole filter, then measure the response to check your work. Then you will modify the 4-pole filter into a 6-pole design. Section 14.5 illustrates the method you will use for specifying the components for an active filter that has a Butterworth response. The method is entitled *Design Guidelines for a Butterworth Filter* and includes a 4-pole filter as an example.

In Part 2, a biquadratic filter, a state-variable filter, and a self-contained IC filter are tested. An advantage of a biquadratic (or "biquad" for short) filter is that it is very simple, consisting of only two integrators, an inverting stage, and a minimum of components. Another advantage of the biquad design is that it can independently control various filter parameters (bandwidth, gain, and center frequency). The biquad design has two outputs – a band-pass output and a low-pass output. This feature is useful in some applications, although it is not possible to optimize both responses. Figure 14-1 shows a basic biquad filter. For this design, $R_{3A} = R_{3B}$ and $R_{4A} = R_{4B}$. Based on these component designations, design equations for the band-pass output are as follows:

$$A_v = -\frac{R_2}{R_1} \qquad Q = \frac{R_2}{R_3} \qquad f_0 = \frac{1}{2\pi R_3 C} \qquad BW = \frac{1}{2\pi R_2 C}$$

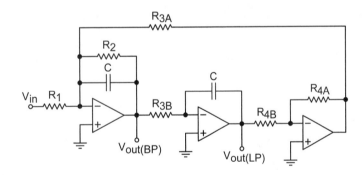

Figure 14-1: A Basic Biquad Filter

14.2 Reading

Floyd, *Electronic Devices, 8th Edition*, Chapter 15

14.3 Key Objectives

Part 1: Calculate and measure the response of a 4-pole Butterworth filter and change the circuit to a 6-pole filter.

Part 2: Test biquadratic, state-variable, and IC filters. Modify the design of the biquad and report on the results of the modification.

14.4 Multisim Files

Part 1: *Devices_Exp_14_Part_01*

Part 2: *Devices_Exp_14_Part_02a*, *Devices_Exp_14_Part_02b*, and *Devices_Exp_14_Part_02c*

14.5 Design Guidelines for a Butterworth Filter

You can design your own Butterworth low-pass or high-pass active filter by using the following guidelines:

1) Determine the number of poles necessary based on the required roll-off rate. Choose an even number, as an odd number will require the same number of op-amps as the next highest even number. For example, if the required roll-off is –40 dB/decade, specify a two-pole filter.

2) Choose R and C values for the desired cutoff frequency (R_A, R_B, C_A, and C_B). For best results, choose resistors between 1 kΩ and 100 kΩ. The values chosen should satisfy the cutoff frequency as given by the equation: $f_c = 1/2\pi RC$.

3) Choose resistors R_f and R_i that give the gains for each section according to the values listed in Table 14-1. The gain is controlled only by R_f and R_i. Solving the closed-loop gain of a noninverting amplifier gives the equation for R_f in terms of R_i: $R_f = (A_v - 1)R_i$.

Table 14-1: Section Gains for Butterworth Low-Pass and High-Pass Filters

Poles	Gain Required		
	Section 1	Section 2	Section 3
2	1.586		
4	1.152	2.235	
6	1.068	1.586	2.483

Example: A low-pass Butterworth filter with a roll-off of approximately –80 dB/decade and a cutoff frequency of 2.0 kHz is required. Specify the components.

Step 1: Determine the number of poles required. Since the design statement specifies a roll-off rate of approximately –80 dB/decade, it requires a four-pole (two-section) filter.

Step 2: Choose R and C. Try C as 0.01 μF and compute R. Computed R = 7.96 kΩ. Since the nearest standard value is 8.2 kΩ, choose C = 0.01 μF and R = 8.2 kΩ.

Step 3: Determine the gain required for each section and specify R_f and R_i. From Table 14-1, the gain of section 1 is required to be 1.152 and the gain of section 2 is required to be 2.235. Choose resistors that will give these gains for a noninverting amplifier. The choices are determined by again considering standard values and are shown on the completed schematic in Figure 14-2.

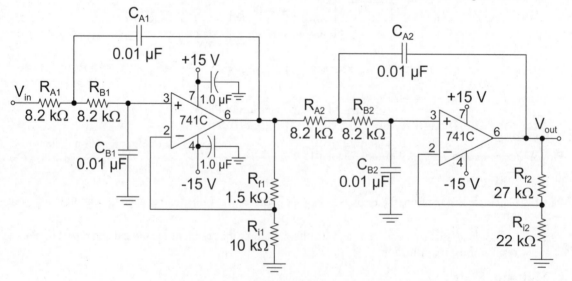

Figure 14-2: Example of Butterworth 4-Pole Filter

Part 1: Low-Pass Butterworth Filter

14.6 Calculating and Measuring Parameters for a 4-Pole Filter

1) Open the Multisim file *Devices_Exp_14_Part_01*. This circuit contains the low-pass filter in Figure 14-2.

2) Use the values in the circuit to calculate the cutoff frequency, f_c (which, by definition, is the frequency at which the gain has dropped −3 dB.) Enter the calculated value in Table 14-2.

3) Calculate the gain at midband, assuming that the resistor values are as given in Figure 14-2. Note that the overall midband gain is the product of the gains for each stage. Express the gain in dB ($A_v(\text{dB}) = 20 \log A_v$). Record the calculated value in the second row of Table 14-2.

4) Calculate the gain at a frequency that is $\sqrt{10}$ beyond the cutoff frequency (a factor of 3.16). This frequency is the frequency at which the response is −40 dB from the cutoff frequency for a 4-pole filter (can you figure out why?). Express the gain at this frequency in dB in Table 14-2. This gives you three points on the response curve to predict (midband, cutoff, and beyond cutoff).

Table 14-2: 4-Pole Butterworth Filter Calculated and Measured Values

Parameter	Calculated Value	Measured Value
f_c		
$A_{v(\text{midband})}$		
$A_{v(\text{X3.16})}$		

5) Connect the Bode plotter so that **IN**(+) connects directly to the source and **OUT**(+) connects to the output (pin 6 of U_2). The (−) inputs should connect to ground. Adjust the controls as shown in Table 14-3:

Table 14-3: Low-Pass Butterworth Filter Bode Plotter Settings

Mode	Horizontal	Vertical
Magnitude	Log F: 20 kHz I: 20 Hz	Log F: 20 dB I: −50 dB

Note that the default resolution is 100. You can increase the resolution by clicking the **Set...** button, but it is not necessary.

6) Activate the circuit and measure the parameters listed in Table 14-2. When you wish to make a reading, you should use the cursor on the left side of the display. Right-click the triangle at the top and enter the independent parameter such as a frequency to obtain the best reading.

14.7 Extending the Design to a 6-Pole Butterworth Filter

1) Use the design guidelines in Table 14-1 to calculate the gain-setting resistor values for a 6-pole low-pass filter with approximately the same f_c (2 kHz). Figure 14-3 shows the basic filter without the gain-setting resistor values.

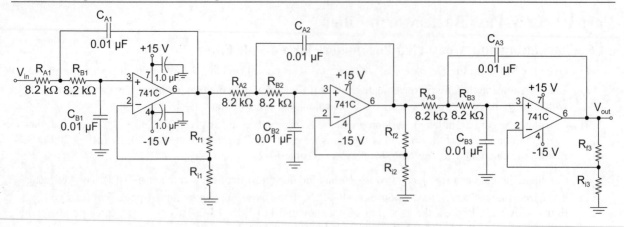

Figure 14-3: A 6-Pole Butterworth Filter

2) Complete the design by writing in the values you selected for the gain-setting resistors. *Hint*: A good starting point is to aim for R_i to be about 10 kΩ. For the first and third sections, this will give you standard resistor values for R_f with only a small error.

3) In the Multisim file, copy and paste the last section of the 4-pole filter to the right side of the circuit and connect this section to the last section.

4) Change the gain-setting resistors in the file to your calculated values.

5) Set up the Bode plotter as before and observe the circuit.

6) Summarize the circuit operation with specific results.

Conclusions for Part 1

Questions for Part 1

1) How would you convert the filter in Figure 14-1 to a high-pass filter?

2) Explain how you would measure the critical frequency using an oscilloscope.

Part 2:   Other Filters

14.8 A Biquadratic Filter

1) Open the Multisim file *Devices_Exp_14_Part_02a*. The circuit is a biquadratic filter that has two outputs – a band-pass output and a low-pass output. In this first circuit, the band-pass filter is optimized first, so the roll-off for the low-pass filter will be as smooth as that for filter in Part 1. By adding another op-amp and more components, a high-pass filter can be constructed.

2) Connect the Bode plotter so that the **IN(+)** is connected directly to the source and **OUT(+)** is connected to the band-pass output (output of U_1). The (−) inputs should connect to ground. Adjust the controls as shown in Table 14-4:

Table 14-4: Biquadratic Filter Bode Plotter Settings

Mode	Horizontal	Vertical
Magnitude	Log F: 10 kHz I: 200 Hz	Log F: 20 dB I: −20 dB

Note that the default resolution is 100. You can increase the resolution by clicking the **Set...** button, but it is not necessary.

3) Using the equations given in the Introduction, calculate the parameters listed in Table 14-5. Express the gain in dB.

4) Measure and record the parameters listed in Table 14-5. Right-click the small triangle on the cursor to move it to a precise position for a measurement. For example, to find the peak, click **Go to next Y_MAX=>**. For Q, the measured value is the measured BW divided by the measured f_0.

Table 14-5: Biquadratic Filter Calculated and Measured Values

Parameter	Calculated Value	Measured Value
A_v		
Q		
f_0		
BW		

5) Using the Bode plotter, observe the LP (low-pass) output. Explain why the response is not particularly desirable for a low-pass filter.

Observations:

6) Think about how to change values in the circuit to optimize the low-pass response. Try to get a flat response out to the low-frequency roll-off. Then make the change or changes and measure the response. Discuss the changes you made to the circuit and how they affected the response.

Changes made to optimize the low-frequency response:

14.9 A State-Variable Filter

State-variable filters are a little more complex than biquad filters, but offer the advantage of a high-pass output. As in the case of the biquad filter, it is not possible to optimize all three responses.

1) Open the Multisim file *Devices_Exp_14_Part_02b*, which is a state-variable filter. This circuit is a compromise with a fairly low Q.

2) Connect the Bode plotter so that the **IN(+)** is connected directly to the source and **OUT(+)** is connected to the band-pass output (output of U_2). The ($-$) inputs should connect to ground. Adjust the controls as shown in Table 14-6:

Table 14-6: State-Variable Filter Bode Plotter Settings

Mode	Horizontal	Vertical
Magnitude	Log F: 50 kHz I: 20 Hz	Log F: 10 dB I: -20 dB

3) Using the equations given here, calculate the parameters listed in Table 14-7 and enter your calculated values in the table. These equations are for the band-pass output.

$$f_0 = 1/(2\pi R_4 C_1)$$
$$Q = (R_5/R_6 + 1) / 3$$
$$BW = f_0 / Q$$

4) Measure and record the parameters listed in Table 14-7 for the band-pass filter. Right-click the small triangle on the cursor to move it to a precise position for a measurement. For example, to find the peak, click **Go to next Y_MAX=>**. For Q, the measured value is the measured BW divided by the measured f_0.

Table 14-7: State Variable Filter Calculated and Measured Values

Parameter	Calculated Value	Measured Value
f_0		
Q		
BW		

5) Observe the low-pass and high-pass outputs, and record your observations. In particular, notice the gain in the main part of the response and the roll-off rate.

14.10 An IC Active Filter

Active filters are available in an IC package that allows the designer flexibility in choosing the type of response, number of poles, etc. One example is the MAX274 IC, which has four independent sections that can be cascaded for higher-order filters. The filters can be implemented with no external capacitors. Because the design equations are somewhat involved, Maxim recommends using their special software to choose the resistor values. A low-pass 8-pole filter is shown in the Multisim file *Devices Exp_14_Part_02c*. Using the methods you have learned, investigate the response as calculated by Multisim. Complete Table 14-8, listing the decibel attenuation (in dB) and the angular response. Summarize your observations.

Table 14-8: IC Filter Attenuation and Phase Shift Measured Values

Frequency	Attenuation	Phase Shift
1 kHz		
2 kHz		
5 kHz		
10 kHz		
20 kHz		
50 kHz		
100 kHz		

Conclusions for Part 2

Questions for Part 2

1) For a basic biquad filter, how would you make the band-pass center frequency variable?

2) What effect does a frequency change have on the Q for a biquad filter?

3) How would you modify the state-variable filter to have a more optimum response for the low-pass and high-pass outputs? What will happen to the band-pass filter as a result?

Devices Experiment 15 - Oscillators

15.1 Introduction

There are two basic types of oscillators - feedback (regenerative) oscillators and relaxation oscillators. A feedback oscillator is basically an amplifier that generates a continuous alternating voltage by feeding a portion of the output signal back to the input in the proper phase. A relaxation oscillator generally uses an *RC* circuit to produce a nonsinusoidal output.

In Part 1 of this experiment, you will investigate a Wien-bridge oscillator. This circuit is a popular feedback oscillator, which is used to generate high-quality sinusoidal waveforms. A portion of the output is fed back to the input in the proper amplitude and phase to reinforce the signal. This type of feedback is called *positive* feedback. For the standard Wien-bridge, the feedback network returns 1/3 of the output signal to the noninverting input. Therefore, the amplifier must provide a gain of exactly 3 to overcome this attenuation and prevent oscillations from dying out. Gain control is accomplished by *negative* feedback. Negative feedback is provided by a FET that is able to adjust gain automatically to maintain the required gain of 3. Another type of feedback oscillator is the Colpitts oscillator that uses two capacitors in the tank circuit. This oscillator taps off a small portion of the output signal and returns it to the input to sustain the oscillations. You will also investigate this oscillator in Part 1.

Part 2 introduces the versatile 555 timer chip, one of the first ICs of its type and still one of the most popular timers. Its operation as an astable multivibrator is both simple and reliable. You will investigate the basic astable multivibrator and observe the effect of modifications to the basic circuit. You will then combine the timer circuit with the Colpitts oscillator from Part 1 to form an "on-off" version of an Amplitude Shift Keying (ASK) circuit, and you will specify and test the resonant components to change the Colpitts oscillator to a Hartley oscillator.

15.2 Reading

Floyd, *Electronic Devices, 8th Edition*, Chapter 16

15.3 Key Objectives

Part 1: Measure parameters for a Wien-bridge oscillator and a Colpitts oscillator. Use a spectrum analyzer to determine the first harmonic distortion from the Colpitts oscillator.

Part 2: Test several relaxation oscillators, including a 555 timer set up as an astable multivibrator, and a circuit that uses a 555 timer to modulate a Colpitts and a Hartley oscillator.

15.4 Multisim Files

Part 1: *Devices_Exp_15_Part_01a* and *Devices_Exp_15_Part_01b*

Part 2: *Devices_Exp_15_Part_02a* through *Devices_Exp_15_Part_02d*

Part 1:   Feedback Oscillators

15.5 The Wien-Bridge Oscillator

1) Open the Multisim file *Devices_Exp_15_Part_01a*. This circuit is a Wien-bridge oscillator, which is capable of generating high-quality sine waves. Calculate the expected output frequency from $f_r = 1/2\pi R_1 C_1$. Enter the calculated f_r in Table 15-1.

2) Connect the Tektronix oscilloscope to view the waveform. Activate the circuit. You will need to adjust the value of R_f using the "A" key to get oscillations to start. Set it for a stable output with a nice sine wave output. Measure the frequency and enter it in Table 15-1.

Table 15-1: Wien-Bridge Output Frequency Calculated and Measured Values

	Computed	Measured
f_r		

3) Measure the peak-to-peak output voltage, $V_{out(pp)}$. Then measure the peak-to-peak positive and negative feedback voltages, $V_{(+)(pp)}$ and $V_{(-)(pp)}$. Measure the dc voltage on the gate of the FET (V_G) using the DMM (it will take a few seconds for the voltage to stabilize). Record all voltages in Table 15-2.

Table 15-2: Wien-Bridge Oscillator Measured Voltages

$V_{out(pp)}$ (Pin 6)	$V_{(+)(pp)}$ (Pin 3)	$V_{(-)(pp)}$ (Pin 2)	V_G

4) Use two channels and observe the phase relationship of the waveforms. What is the phase shift from the output voltage to the positive feedback voltage?

Phase Shift:

5) Add a second diode in series with the first one between the output and the gate of the FET (See Figure 15-1). Readjust R_f for a good sine wave. Measure the voltages as before and record in Table 15-3.

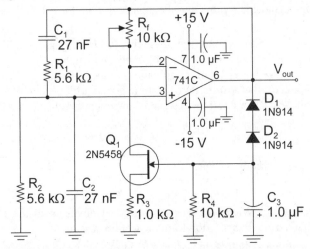

Figure 15-1: Modified Wien-Bridge Oscillator Circuit

Table 15-3: Wien-Bridge Oscillator Measured Voltages

$V_{out(pp)}$ (Pin 6)	$V_{(+)(pp)}$ (Pin 3)	$V_{(-)(pp)}$ (Pin 2)	V_G

15.6 The Colpitts Oscillator

1) Open the Multisim file *Devices_Exp_15_Part_01b*. This circuit is a Colpitts oscillator that uses a capacitive voltage divider to tap off a feedback signal. Calculate and record the two parameters listed in Table 15-4.

   a. The oscillation frequency f_r is given by

   $$f_r = \frac{1}{2\pi\sqrt{LC}}$$

   The capacitance C in this equation is the series combination of C_1 and C_2.

   b. The feedback fraction, B, is the fraction of the output that is fed back to the base of the transistor. Use the capacitive voltage-divider formula to calculate the feedback fraction.

2) Activate the circuit. It will take a few seconds for oscillations to begin. Measure the frequency of the output oscillation and record it as the measured frequency in Table 15-4.

3) Measure the peak-to-peak signal voltage at the base, V_b, and the peak-to-peak output voltage, V_{out}. The ratio V_b / V_{out} is the feedback fraction, B. Record this in Table 15-4.

Table 15-4: Colpitts Oscillator Calculated and Measured Values

Parameter	Calculated Value	Measured Value
f_r		
B (feedback fraction)		

4) Connect the spectrum analyzer to the output and set it up to view the fundamental and first harmonic. Determine the amplitude of the first harmonic relative to the fundamental. Summarize your observations.

Conclusions for Part 1

Questions for Part 1

1) For the Wien-bridge, why do you think the output signal increased with a second diode in series?

2) What is the purpose of C_3 in the Wien-bridge circuit in Figure 15-1?

3) From your observation with the spectrum analyzer for the Colpitts oscillator, what conclusion can you make about the distortion in the output?

Part 2: Relaxation Oscillators

15.7 A Square Wave and Triangle Wave Oscillator

1) Open the Multisim file *Devices_Exp_15_Part_02a*. This circuit is essentially putting together two circuits you have already seen. The first circuit is the Schmitt trigger comparator from Experiment 12, modified with an *RC* circuit on the inverting input; the second is the integrator introduced in the same experiment. By putting these elements together, a simple square wave and triangle wave generator is constructed. Connect the oscilloscope so that channel 1 monitors the square wave and channel 2 monitors the triangle. Activate the circuit and adjust the scope to view the input and output. Summarize your observations:

2) The circuit in step 1 is a good signal generator, but it can be simplified and produce the same result! Instead of a separate *RC* circuit to develop the time constant, you can use feedback from the integrator output. The modified circuit is in the Multisim file *Devices_Exp_15_Part_02b*. Open this file and activate the circuit. (Notice that component numbers were changed to match Floyd's text.) Compare the measured frequency

with the frequency f_r predicted from the equation $f_r = \dfrac{1}{4R_1C}\left(\dfrac{R_2}{R_3}\right)$.

Observations:

15.8 The 555 Timer as an Astable Multivibrator

1) Open the Multisim file *Devices Exp_15_part_2c*. The circuit is a basic astable multivibrator using a 555 timer. Calculate and then measure the parameters listed in Table 15-5 for the circuit with both switches open. Equations for the basic astable multivibrator are

$$f_r = \frac{1.44}{(R_1 + 2R_2)C_{ext}} \quad t_H = 0.694(R_1 + R_2)C_{ext} \quad t_L = 0.694R_2C_{ext} \quad \text{Duty cycle} = \frac{(R_1 + R_2)}{(R_1 + 2R_2)}100\%$$

Table 15-5: 555 Timer Astable Multivibrator Calculated and Measured Values

Parameter	Calculated Value	Measured Value
f_r	3.31 kHz	3.31 kHz
t_H	198 μs	196 μs
t_L	103 μs	106 μs
Duty cycle	0.657	0.649

2) Use the space bar to close J_1 and observe what happens. Summarize your observations.

3) Use the "A" key to close J_2 and observe what happens. Summarize your observations.

4) Open J_2 and view the waveform on the top of C_{ext} with Channel 2. Summarize your observations.

5) Assume you needed to have a 50% duty cycle circuit with a frequency of about 10 kHz. Calculate a value of C_{ext} that will meet this specification using the circuit from the previous step. Then change the capacitor to your value and measure the frequency.

$C_{ext} =$ $f_r =$

15.9 Combining the 555 Timer and Colpitts Oscillator

1) Open the Multisim file *Devices_Exp_15_Part_02d*. This circuit uses an astable 555 circuit to gate a Colpitts oscillator on and off. This produces "bursts" of the sine wave from the Colpitts oscillator, forming

an on-off keying version of an ASK test generator as described in Floyd's text. The scope is set up to go, so that you can view the signals. A few values are different from the text. Describe the observed signals and measure the frequency of Colpitts oscillator.

Observations:

2) Assume your boss comes to you and says, "The Company just bought a new coil winding machine. We can wind coils and they are cheaper than the capacitors we buy. Can you change the oscillator to a Hartley oscillator? We need you to change only the resonant circuit, which must oscillate at the exact same frequency and we have lots of 10 nF capacitors, so please use one of these in your design." Draw the resonant tank circuit you will use and test it by modifying the Multisim design.

Observations:

Conclusions for Part 2

Questions for Part 2

1) In the triangle wave generator, if you lower the frequency by increasing R_1, will there be any effect on the amplitude of either of the two outputs? Explain your answer.

2) Why did inserting the diode in the 555 astable circuit affect the frequency and the duty cycle of the output?

3) Does the power supply setting have an effect on the frequency from the 555 timer? Why or why not?